NETWORK MANAGEMENT IN WIRED AND WIRELESS NETWORKS

THE KLUWER INTERNATIONAL SERIES IN ENGINEERING AND COMPUTER SCIENCE

NETWORK MANAGEMENT IN WIRED AND WIRELESS NETWORKS

by

Tejinder S. Randhawa
Acterna, Canada

Stephen Hardy
Simon Fraser University, Canada

KLUWER ACADEMIC PUBLISHERS
Boston / Dordrecht / London

ISBN 978-1-4419-4931-8

Distributors for North, Central and South America:
Kluwer Academic Publishers
101 Philip Drive
Assinippi Park
Norwell, Massachusetts 02061 USA
Telephone (781) 871-6600
Fax (781) 681-9045
E-Mail < kluwer@wkap.com>

Distributors for all other countries:
Kluwer Academic Publishers Group
Distribution Centre
Post Office Box 322
3300 AH Dordrecht, THE NETHERLANDS
Telephone 31 78 6392 392
Fax 31 78 6546 474
E-Mail < services@wkap.nl>

Electronic Services < http://www.wkap.nl>

Library of Congress Cataloging-in-Publication Data

A C.I.P. Catalogue record for this book is available
from the Library of Congress.

Printed on acid-free paper.

Printed in the United States of America

Table of Contents

List of Figures

List of Tables

Preface

Until recently, the focus of the standards organizations, as well as the telecommunications industry, has been switching and transmission technology. Numerous protocols, devices and transmission technologies have been released in the market place to offer high-speed services. Now that these new technologies are being integrated into the networks of service providers, an effective management of these heterogeneous and complex networks is being identified as the missing link. There is a general acceptance of the understanding that even though these networks are composed of high capacity and intelligent NEs (Network Elements) and transport mechanisms, they still, at times, experience periods of performance degradation, congestion and even failures, leading to violations of SLAs (Service Level Agreements). This may, consequently, result in lost revenues for the service providers and, perhaps, lost productivity for the end user. An effective management strategy is, therefore, needed to anticipate, detect and subsequently overcome these problems. Network management is thus being given the spotlight it deserves.

Most of the existing books on this subject elaborate on the definitions, functional requirements and architectures of network management. This book, however, has been written with a focus on describing various *methods* of network management. With an emphasis on the performance management aspects of telecommunication networks, this book addresses various open issues related to performance monitoring, performance management and performance control. The book makes two key contributions. Firstly it covers the performance management aspects of broadband wired and wireless cellular networks in an integrated fashion. Secondly it highlights the role of performance management in assisting network control procedures to achieve the end objectives. The end objectives are the high return on investment for the service providers and an acceptable QoS for the end user.

The first half of the book provides details of performance monitoring in broadband wired and wireless cellular networks. Various parameters that characterize the performance of these networks have been identified. Chapter 3 elaborates on the role of signaling networks in determining network performance, especially of GSM/GPRS cellular networks.

The second half of the book presents and analyses methods of performance optimization in broadband wired and wireless cellular networks. The role of management systems in the network control of these networks at the packet level, call level and network level is described. Chapter 4 elaborates on an application of a monitoring system in facilitating congestion control of ATM/B-ISDN networks by estimating and predicting multiplexed traffic. Chapters 6 and 7 describe schemes of bandwidth optimization. These schemes optimize throughput and QoS in multi-service broadband wired and wireless cellular networks.

Tejinder S. Randhawa

Stephen Hardy

September 2001

I dedicate this book to Avineet, Jesse, my parents, and to the memory of my brother, Ranjit S. Randhawa.

--Tejinder S. Randhawa

I dedicate this book to the memory of my father, Robert M. Hardy, and to those who have helped me in this journey.

--Stephen Hardy

Chapter 1

Introduction

As the telecommunication technology continues to develop and support an increasing range of services, the management of this technology and the way it integrates into the network operator's infrastructure will become critical to the success of using such systems. The management requirements of these networks, therefore, need to be analyzed and well understood in order to identify the most optimal solutions. A major step towards this direction has been the development of Telecommunications Management Network (TMN) standards by ITU-T (formerly known as CCITT) [58], [60]. This TMN standardization process has resulted in the identification of informational, architectural, and functional requirements for the management of telecommunication networks. These network management requirements have been classified into the following main categories:

- Fault Management
- Configuration Management
- Accounting Management
- Performance Management
- Security Management

This work deals primarily with Performance Management, both monitoring (detection) and control (resolution). Of all the above mentioned tasks, the task of performance management is that of pursuing high-level management objectives. These high level objectives can be grouped into two main classes. The first class deals with providing network services that meet the needs of customer applications such as service reliability and Quality-of-Service (QoS) guarantees. The second class deals with defining resource allocation strategies that provide benefits for the service provider. As an example, controlling call blocking rates would belong to the first class of management objectives while pursuing high resource utilization and favoring one type of traffic over others would fall into the second category. The first class of management objectives favors increasing the resources allocated to each call while the second class focuses on achieving a high level of resource utilization. These are conflicting requirements which have to be balanced and optimized.

The focus of this work, therefore, is on introducing and discussing aspects, issues and goals of performance management in the telecommunication networks.

Chapter 2 deals with performance management in broadband networks, with particular focus on performance monitoring. Performance monitoring functions and requirements are discussed in the context of T1/T3, SDH/SONET, ATM/B-ISDN, and Frame Relay based networks. An introduction to transmission quality assurance and traffic management is also given.

Chapter 3 deals with performance management in wireless cellular networks, again with a focus on performance monitoring. A performance management system for GSM/GPRS networks is introduced, which makes use of signaling data to estimate performance parameters. Key performance parameters associated with GSM/GPRS networks are identified and the use of signaling data in deducing these parameters is detailed.

The remainder of this work presents methods of performance control of both broadband wired and cellular wireless networks. A comprehensive framework of QoS and throughput optimization for both types of networks, is proposed.

Chapter 4 discusses estimation and prediction of muliplexed traffic. Specifically the feasibility of estimating and predicting multiplexed VBR traffic in a high-speed packet-switched network such as ATM is evaluated. The multiplexed VBR traffic is characterized and appropriate filters are proposed for estimation and prediction. Simulation results, demonstrating the feasibility of predicting multiplexed VBR traffic, are presented.

Chapter 5 introduces approaches to channel access control. Priority based channel access control policies are proposed and analyzed using multi-dimensional Markov chain analysis. Computational results are presented that demonstrate the benefit of these techniques as compared to the conventional techniques such as CS, CP and PS.

Chapter 6 discusses bandwidth optimization, and a bandwidth optimization procedure, in broadband networks. A framework of QoS evaluation and capacity optimization in multi-service broadband networks is presented. The QoS and throughput for the entire network is computed and an optimal channel allocation plan is determined.

Chapter 7 introduces approaches to capacity optimization in cellular wireless networks, and discusses quantitative techniques for both cell capacity estimation and cell level QoS evaluation. The analysis of Chapters 5 and 6 is extended to cellular networks by taking into consideration user mobility and soft capacity in CDMA.

1.1 Motivation for Performance Control

The demand for multimedia services such as video-conferencing and video-phones is growing rapidly. Also growing is the desire to receive these services anytime and anywhere. Related services are being integrated onto a single network rather than being offered on separate uncoordinated networks. The existing voice-only circuit-switched networks are, therefore, fast evolving into multi-service & multi-resource broadband networks composed of circuit-switched, packet-switched and cellular network domains, as illustrated in Figure 1-1. Supporting a matrix of diverse services with an acceptable QoS (Quality of Service) for a large number of subscribers is essential for the economic viability of these networks. The key to supporting a large number of subscribers with an acceptable QoS is the efficient utilization of bandwidth. In wireless networks the bandwidth is always at a premium. Even on the wired side, at currently available user access speeds, it does not take too many connections to saturate a link that operates in Gbps (Giga bits per second) range. Emerging applications are thus changing the premise that network bandwidth will ever be an unlimited resource. Integrated services will need to share resources both for functionality as well as to decrease cost. Sophisticated bandwidth optimization procedures, therefore, need to be incorporated to maximize network throughput or revenue, and achieve the desired QoS.

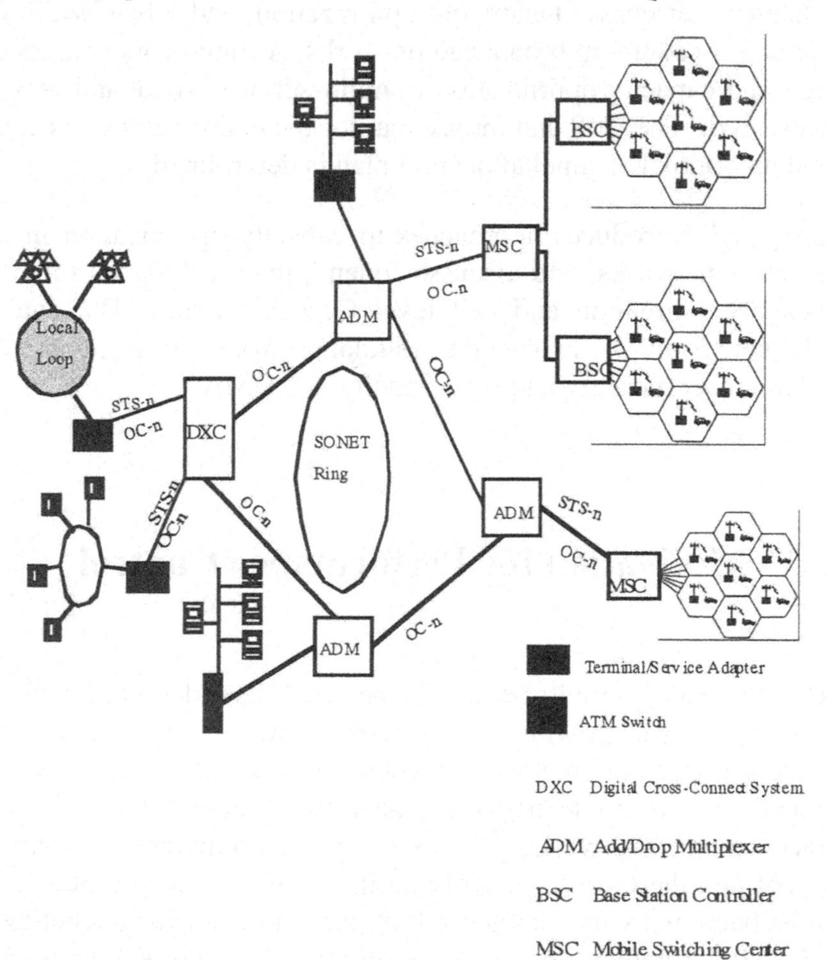

Figure. 1-1. A Multi-Service Multi-Resource Broadband Network.

Among the current trends in the telecommunication industry, SDH/ SONET (Synchronous Digital Hierarchy/ Synchronous Optical Networks) has emerged as the transmission technology of choice that offers over 2.5 Gbps capacity, high flexibility, and high resilience to transmission errors [4]. Protocols are being developed and deployed to transport broadband packet data over SDH/SONET, in the form of ATM (Asynchronous Transfer Mode) over SDH/SONET or IP (Internet Protocol) over ATM over SDH/SONET, that enable service providers to offer emerging broadband ATM services, while continuing to support existing telephone services and circuit-switched broadband services. Typically, some bandwidth channels of an SDH/SONET link are used to support entities such as permanent ATM VPs (Virtual Paths), permanent ATM VCs (Virtual Channels), or trunks for the public switched telephone networks etc.; whereas the remaining channels could be used for switched telephone calls or switched ATM VCs. From network management perspectives the challenge is to determine the optimum sizes of the SDH/SONET 'pipes' or number of channels allocated to the services sharing the transmission medium.

On the wireless side, the demand for services is growing at an unprecedented rate. Although cellular phones and low bit rate data continue to constitute the larger portion of this demand, the wide-band services such as video-phones and video-conferencing are also beginning to generate enough interest. CDMA (Code Division Multiple Access) based 3G (Third Generation) wireless networks are expected to provide bit rates as high as 2Mbps per user that will further facilitate the support for such services. The next generation cellular networks are thus also expected to evolve into multi-service networks supporting narrow-band as well as wide-band services. The main limitation to supporting a large number of subscribers, however, continues to be the scarcity of the available radio frequency spectrum. The number of subscribers per service per unit area that can be supported at some minimum QoS is an important parameter. Spectrally efficient bandwidth optimization procedures can contribute significantly to achieve the desired throughput. From network management perspectives, the challenge is to determine the optimum portions of the available spectrum that should be reserved for each offered service at each cell site to cater to the corresponding traffic load.

In a very general context, the problem of bandwidth optimization is stated as under:

Given a network topology, a portion of the bandwidth of the physical links, both wired as well as wireless, is reserved in an end-to-end fashion to support multiple services or classes of users. The objective is to partition the bandwidth in such a way that the overall network throughput or revenue is maximized, while ensuring the target QoS guarantees.

Accomplishing unified bandwidth optimization in such heterogeneous networks, which are composed of diverse switching and transmission technologies would require the implementation of control mechanisms at the packet level, the call level, and the network level. Packet level control is required to provide congestion free high-speed transmission of packets in the packet-switched portions of the network. A packet level controller monitors the flow of packets along the network links and takes actions to avoid violations of packet level QoS guarantees such as packet loss rate, packet delay and packet delay variations. The call level control, on the other hand, works at the traffic granularity of a call. It allows a call to be accepted or rejected based on the availability of resources or some explicit access control criteria. The objective at this level is to prevent network congestion while assuring call level QoS guarantees such as call blocking probability, call dropping probability and call set-up delay. Finally, the network level control determines a network wide optimal bandwidth allocation plan that maximizes the overall network throughput or revenue, while assuring that the overall QoS is within the prescribed bounds. The computed parameters are passed on to the call level as well as packet level controllers. These three levels of control are required to operate in a nested fashion to achieve end-to-end performance objectives.

In chapters 4 to 7 of this work, some open issues associated with the said control mechanisms are identified and resolved. Firstly, we address the issue of congestion control in packet-switched networks carrying VBR traffic. We demonstrate, using synthetic as well as empirical traffic traces, the feasibility of predicting multiplexed VBR video traffic. This capability can significantly enhance the effectiveness of packet level control in preventing congestion. Secondly, we address the issue of preferential call admission control by proposing and analyzing two priority based channel access control policies. These policies give preferential treatment to high priority services at the expense of low priority services. In one of the proposed policies a low priority call can be preempted to make room for a high priority call; while in the other policy the low priority arrivals are throttled as the network congestion increases. We demonstrate using computational results that these policies perform better than conventional policies. The performance is evaluated in terms of call level QoS parameters, primarily the call blocking probability and the call dropping probability. Thirdly, we develop a robust yet flexible bandwidth optimization framework for multi-service broadband networks. A heuristic based approach is proposed to optimize network wide bandwidth allocations to competing services or classes of user. Lastly, we extend this framework to multimedia wireless cellular networks. The adaptation of channel access control policies, mostly studied under multi-service circuit-switched demand access, into cellular domains is accomplished by taking into consideration two key aspects of cellular networks. One of these aspects is the user mobility, while the other is the soft cell capacity in CDMA networks. The user mobility is incorporated by assuming that the call arrivals to the base stations are Poisson with exponential channel holding times. The soft capacity of the CDMA cells, on the other hand, is handled by quantifying the statistical multiplexing gain of multiple video and voice users in a cell in terms of bit error rate requirements.

Based on the solutions, mentioned above, a comprehensive and viable framework of QoS and throughput optimization in broadband wired as well as wireless networks, is proposed. The subsequent sections provide further details of the aforementioned contributions and review the published literature that is most relevant to these topics.

1.2 Related Work

1.2.1 Packet Level Control

As mentioned earlier, the main objective of the packet level control is to prevent or alleviate congestion at the packet switches. The subject of congestion control has been studied quite extensively for the conventional low speed packet-switched networks and for which several workable solutions exist. These conventional, reactive based, approaches to congestion control, however, fail to scale well to the extremely high speeds, wide area distances and large volumes of traffic associated with the high speed packet-switched networks such as ATM networks [25][50]. Various preventative based schemes have also been specified for ATM based B-ISDN networks that aim at reducing the possibility of congestion in the network by proper shaping of traffic at the source or by proper allocation of resources. However, keeping in view the statistical and dynamic nature of the broadband traffic, these techniques alone are not sufficient to guarantee that the network will always operate in a non-congested state. VBR video is one such broadband traffic type, and is expected to be dominant in ATM based B-ISDN (Broadband-Integrated Services Digital Networks) networks.

The characteristics of VBR traffic such as high peak-to-mean bit rate ratio and stochastic variations of bit rate with respect to time make bandwidth allocation quite difficult. Choosing peak rate as a criteria would mean under-utilization of the available network bandwidth, whereas, allocating bandwidth based on mean rate would lead to congestion in the network. The effect of a large delay-bandwidth product, as seen in ATM networks, also adds to the complexity and renders conventional reactive based congestion control schemes as highly inefficient. Some over-provisioning coupled with a proactive congestion control scheme that predicts the onset of congestion well in advance is therefore the desired solution for managing such traffic in ATM networks.

In Chapter 4, the feasibility of predicting VBR traffic is evaluated. Such capability can significantly improve the effectiveness of tasks such as preventative congestion control. The scheme involves estimating the current traffic state over small observation intervals and predicting the future traffic (Randhawa and Hardy [37, 38, 39]). Using this capability, the onset of congestion could be forecasted and, subsequently, a preventative control action could be taken ahead of time. The control action may involve sending signals back to the preceding network nodes or to the traffic sources to perform flow control; borrowing bandwidth of some non-real time sources; borrowing bandwidth of sources with less stringent QoS requirements; or sending signals to cause traffic rerouting at the preceding nodes.

An important requirement of the proposed technique is the source traffic model. A number of VBR video source models have been proposed for video conferencing sources employing DCT (Discrete Cosine Transform) coding techniques for intra-frame compression with no motion compensation. These models include Markov Modulated Fluid Flow (MMFF) model [31], Markov Modulated Poisson Process (MMPP) model [45, 47, 48], Markov Modulated Deterministic Arrival model (MMDA) [45, 47, 48], and AutoRegressive model [31]. In the proposed approach (Randhawa and Hardy [37, 38]) an AR modulated model, derived from the one proposed by Maglaris *et al.* [31], is assumed. The first and second order statistics of the traffic at the multiplexer output are derived. These statistics are used for determining the coefficients of time-invariant least mean-square filters that are proposed for estimation and prediction of the aggregate VBR video traffic. The performance of the approach is evaluated using synthetic data.

Broadcast quality video sources, employing MPEG (Motion Pictures Experts Group) based compression, are also considered (Randhawa and Hardy [39]). Time-sequenced adaptive filters are proposed to predict multiplexed traffic from such sources. Predicting MPEG traffic from broadcast video is, however, anything but trivial. Unlike video-conferencing or video-phone sources, broadcast video is a collection of scenes of short durations, with traffic statistics changing considerably at scene boundaries, depending upon the underlying activity in the scene. The frames within a scene, however, are similar and, thus, are correlated because of similarities within the scene. The MPEG based compression of such video results in a highly complex bit-rate process, that is difficult to model. This complexity is primarily caused by two main factors. One is the different statistical properties of I (Intra-frame), P (Predictive), and B (Bidirectional-Predictive) frames of an MPEG trace; and the other is the pseudo-periodic nature of the traffic that depends on its GOP (Group of Pictures) structure. Traditional stochastic models, thus, cannot be used to define this traffic.

In the proposed approach, the traffic from MPEG source(s) is modeled as a concatenation of PAR (Periodic AutoRegressive) cyclostationary processes [17]. A PAR process quite suitably captures the stochastic nature as well as the periodicity of the bit rate process within the scene. Based on this model, the traffic at the multiplexer output is estimated and predicted, using a periodic or time-sequenced adaptive filter [16, 30]. The time sequenced adaptive filters are the simple extensions to the conventional adaptive filters, and are capable of adapting to the inherent cyclo-stationarity of multiplexed MPEG traffic with unknown statistics. The validity of the model and the effectiveness of the predictor is verified using computer simulations, in which a number of half-hour long empirical broadcast quality MPEG-1 traces are used to generate the multiplexor output.

Various other prediction based congestion control schemes have also been proposed in recent years. Notable among these are [25], [47], [50], and [56] where predictability of VBR video traffic from video-conferencing or video-phone source is studied. Yu and Chen [56] compare the performance of a linear predictor based on a Box-Jenkins AR model with a neural network based predictor for predicting traffic from a single VBR source. Skelly *et al.* [47] use a matrix expansion approach for predicting congestion in the network receiving input from a video as well as an image source. Kolarov *et al.* [24], and Tsingotjidis *et al.* [50] use Kalman filter to estimate CBR (Constant Bit Rate) ON-OFF sources or quantized VBR sources. Prediction of an MPEG source, on the other hand, has been evaluated by decomposing MPEG traffic into I, P and B frame series and then predicting these series separately [1, 8, 21, 52]. All of these techniques, consequently, are limited to predicting traffic from a single source only and can only be used in the vicinity of the source. The proposed method, in contrast, aims at predicting the multiplexed traffic and could be used, ad-hoc, at any VP in the network carrying multiplexed VBR traffic.

1.2.2 Call Level Control

Based on the availability of resources, the call level controller accepts or rejects a call. The objective, however, is to minimize the call rejection rate. In order to make a sound call admission control decision, an accurate assessment of the resource demand and availability is needed. For CBR traffic the peak rate is used for this assessment, whereas for VBR traffic the effective bandwidth of the calls, lying somewhere between the peak rate and the mean rate, is used [45]. The effective bandwidth is estimated by considering the long term stochastic characteristics of the individual source traffic streams; the number of sources being statistically multiplexed; and the packet level QoS requirements. An effective bandwidth equal to the peak rate corresponds to a very conservative admission policy drawing no benefit from statistical multiplexing, whereas an effective bandwidth equal to the mean rate may correspond to the expectation that a large number of calls when multiplexed will average out the variations in the traffic. The effective bandwidth thus may not account for the instantaneous fluctuations in the source traffic. Besides, there is always the possibility that the traffic from the independent sources may peak at the same time, causing QoS violations. In high speed packet-switched networks, call level control therefore needs to be used in conjunction with the packet level control (as suggested in the last section) to appropriately optimize the call level and the packet level QoS requirements.

The availability of resources need not be the only criteria for call admission control. It may also be based on some specified access control criteria that regulates the access of different types of demands to the bandwidth. In a multi-service network the resources such as link bandwidth are shared by multiple services or classes of users. Each service or user class may have distinct traffic characteristics, bandwidth requirements and QoS expectations. A policy, therefore, may suggest not to accept a call even if the resources are available but instead hold out for a high paying request. The call level control may thus accept or reject a call based on a prespecified channel access control policy. A number of channel access control policies have been discussed in literature [2, 20, 33, 36]. Notable among these are CS (Complete Sharing), CP (Complete Partitioning) and PS (Partial Sharing). In order to analytically determine the most suitable policy for a link, or to fine tune an existing one, these policies need to be evaluated using an accurate call model. The call model implies traffic characteristics such as call arrival distribution and call holding time distribution. Using the call model, the channel access control policies are compared in terms of QoS parameters, such as call blocking probability and call dropping probability. Based on such analysis a new appropriate policy could be derived or an existing one could be fine tuned. Aein [2] analyzed the aforementioned policies with the assumption that the call arrival process is Poisson; Chandramohan [33] assumed it to be a Markov Modulated Poisson process; and Qian *et al.* [36] assumed non-stationarity in the call arrival process. Here, again, Poisson arrival is assumed, though, the analysis is expanded to include also some priority based channel access control policies, in addition to the aforementioned conventional policies (Randhawa and Hardy [41]). The two proposed priority based policies are named 'PS with Call Dropping' and 'PS with Discouraged Arrivals'. Numerical results are presented that demonstrate that these policies substantially improve the throughput of the higher priority services at the expense of low priority services.

This analysis is further extended to also include FDMA, TDMA and CDMA (Frequency, Time, and Code Division Multiple Access) based cellular networks (Randhawa and hardy [42]). The analysis takes into consideration the two main aspects of cellular networks. One is the user mobility and the other is the soft capacity of the CDMA cells. The adaptation of channel access control policies, mostly studied under circuit-switched demand access, into cellular domains is one of the main contribution of this work. Although such policies have been analyzed under the context of FDMA/TDMA based wireless networks by Epstein and Schwartz [14], a single cell network was assumed. In this work, a multi-cell network with mobile users is considered. In the wired broadband telecommunication networks, the call model sufficiently captures the traffic state of the network. Traffic parameters such as average call arrival rate, average call holding time and the effective bandwidth of the calls are used for estimating bandwidth requirements. In cellular networks, however, the user mobility significantly impacts the network traffic characteristics and needs to be taken into consideration. The obvious effect of the subscriber mobility is the call handover. As soon as the mobile moves out of a cell, the channels occupied in the old cell are released and the new cell then caters to the resource demand of this handover call. The channel occupancy time in a cell is, therefore, usually less than or equal to the total call duration, as the mobile may be moving from cell to cell. The user mobility, thus, needs to be modeled and integrated with the call model for an accurate assessment of the network traffic and bandwidth demand.

The common approaches to quantify user mobility are based on fluid [29], Markovian [22], or gravity [6] models. These models are mostly used for characterizing user mobility in situations where sufficient empirical data is not available. Due to the underlying assumptions about subscriber distribution and subscriber mobility, these models do not always render an accurate representation of the network traffic. In the proposed framework, the traffic model is assumed to be adaptively developed, based on the mobility traces observed from the network. It is a generalized Markovian model where transition probabilities to neighboring cells are estimated empirically through network monitoring. Given a network topology, a matrix of transitions to neighboring cells is developed with each element of the matrix representing the probability of transition from cell n_i to n_j. The call origination process in a cell and the handover call arrival process to that cell are both assumed to be Poisson. The channel occupancy time in a cell as well as the total call duration are assumed to be exponential. Assuming independence among cells, the cells are simply M\M\m\m queues. Based on these assumptions, the channel access control policies are evaluated under the context of a cellular network, using multi-dimensional Markov chain analysis, just as in a multi-service circuit-switched network.

We also develop an analytical framework to address the soft capacity of CDMA cells. Unlike FDMA/TDMA where each user is assigned an available frequency/time slot to communicate, the spread-spectrum based DS-CDMA (Direct-Sequence Code Division Multiple Access) networks assign a unique code to each user. The user information is spread by using the code and transmitted over the shared frequency channel. The same code is used at the receiver's end to de-spread and recover the target signal from the aggregate traffic. As there is no physical limit to the number of codes that could be assigned (provided orthogonality among codes is ensured) the capacity of a CDMA cell also does not have a hard upper limit. The capacity however is interference limited and depends on traffic aggregation and power control. Determination of maximum number of sources that can be aggregated without violating SINR (Signal to Interference and Noise Ratio) is essential for bandwidth optimization purposes. Analytical models are, therefore, developed herein that characterize the channel utilization by the multimedia sources including VBR video. Subsequently, the statistical multiplexing gain *i.e.*, the extent of traffic aggregation that is feasible in a CDMA cell with acceptable SINR, is quantified.

Although CDMA is an active area of research, the focus mostly has been on the physical layer engineering. Notable exceptions include [10], [11], [15], [28], [46], and [55] where traffic control has been in the highlight. Comaniciu and Mandayam [10], Chan [11], and Yang and Geraniotis [55] considered only voice and data sources, and proposed admission control policies that minimize interference and reduce rejection rate of real-time traffic. Evans and Everitt [15], and Shen and Ji [46] developed effective bandwidth based call admission control policies for a general multi-service CDMA network. Larijani *et al.* [28] considered both voice and low bit rate VBR video traffic, and presented an access control strategy involving packet as well as call level control. Only one channel is assumed to be used by each source and the channel activity is either assumed or estimated through monitoring. In contrast, the analytical models developed herein are simple, and include real-time VBR video services, in addition to voice and CBR (Constant Bit Rate) services. Traffic from VBR video sources is modeled as an AR (AutoRegressive) process with a Gaussian distribution. Bursty data is assumed to be handled using multiple parallel channels as suggested in the IS-95-B and cdma2000 standards [24, 34]. The use of multiple parallel channels to handle bursty traffic not only alleviates the inherent problem of variable spreading gain associated with VBR sources in CDMA environments, but also simplifies capacity planning by facilitating capacity utilization at finer granularity. The channel activity parameters are estimated based on the assumptions about the underlying source traffic.

1.2.3 Network Level Control

The objective of the network level control is to optimize the capacity allocations to competing services or classes of users so that the overall network revenue is maximized while assuring the target QoS. A simple heuristic based approach is proposed for this purpose (Randhawa and Hardy [40]). Given the projected traffic load, the channel access control policy at each link (a wired link or a cell site) are evaluated in terms of call level QoS parameters such as call blocking probability and call dropping probability. The most appropriate policy is thereafter selected for each link and cell in the network.

The bandwidth dimensioning problem is not new and has appeared in numerous disguises. It has been studied under the context of dimensioning of circuit-switched networks [20,33]; configuration of logical channels in SONET-based networks [23]; VP distribution in ATM networks [3]; and capacity planning of wireless LANs [43]. The proposed approach is a simple alternative to the integer linear programming based techniques that are also used for similar tasks. The integer linear programming based techniques, however, tend to become computationaly intensive as the network size increases, and the matrix of offered services and corresponding QoS requirements become larger. The proposed approach is incremental and builds upon the existing configuration of the network.

1.2.4 The Proposed Framework

Based on the aforementioned analytical results, a framework of QoS evaluation and bandwidth optimization in broadband wired as well as cellular networks, is developed. Figure 1-2 depicts the functionality of the proposed system as embedded in the network management cycle. The overall functionality includes traffic data collection, traffic forecasting, QoS evaluation and bandwidth optimization. The network is continuously monitored to retrieve the traffic information. The register counts of various OAM (Operations Administration and Maintenance) variables from SDH/SONET digital cross-connects and add-drop multiplexors, BSCs (Base Station Controllers) and MSCs (Mobile Switching Centers) are monitored. The HLR (Home Location Registry) and VLR (Visitor Location Registry) databases are also periodically queried to develop a comprehensive knowledge of user mobility in the cellular portions of the network. The instantaneous packet rate of each ATM VC is measured by monitoring volume of ingress/outgress packets at the ATM switches.

Multi-service demands per source-destination pair are projected. If there is no significant projected change in the offered load, then no action needs to be taken. If the load is expected to change significantly, then the QoS for the projected load is computed by taking into account the channel access control policies implemented in the network. When there is a predicted violation of the QoS guarantees, the system determines a channel allocation plan that can optimally accommodate this projected increase in the load. The suggested changes to channel allocation plan are then transmitted to the network switches by the network configuration manager or the provisioner. The resulting plan is held constant until the next run. The proposed system is a network management solution that compliments the real-time call admission control and packet level control procedures already implemented in the underlying network.

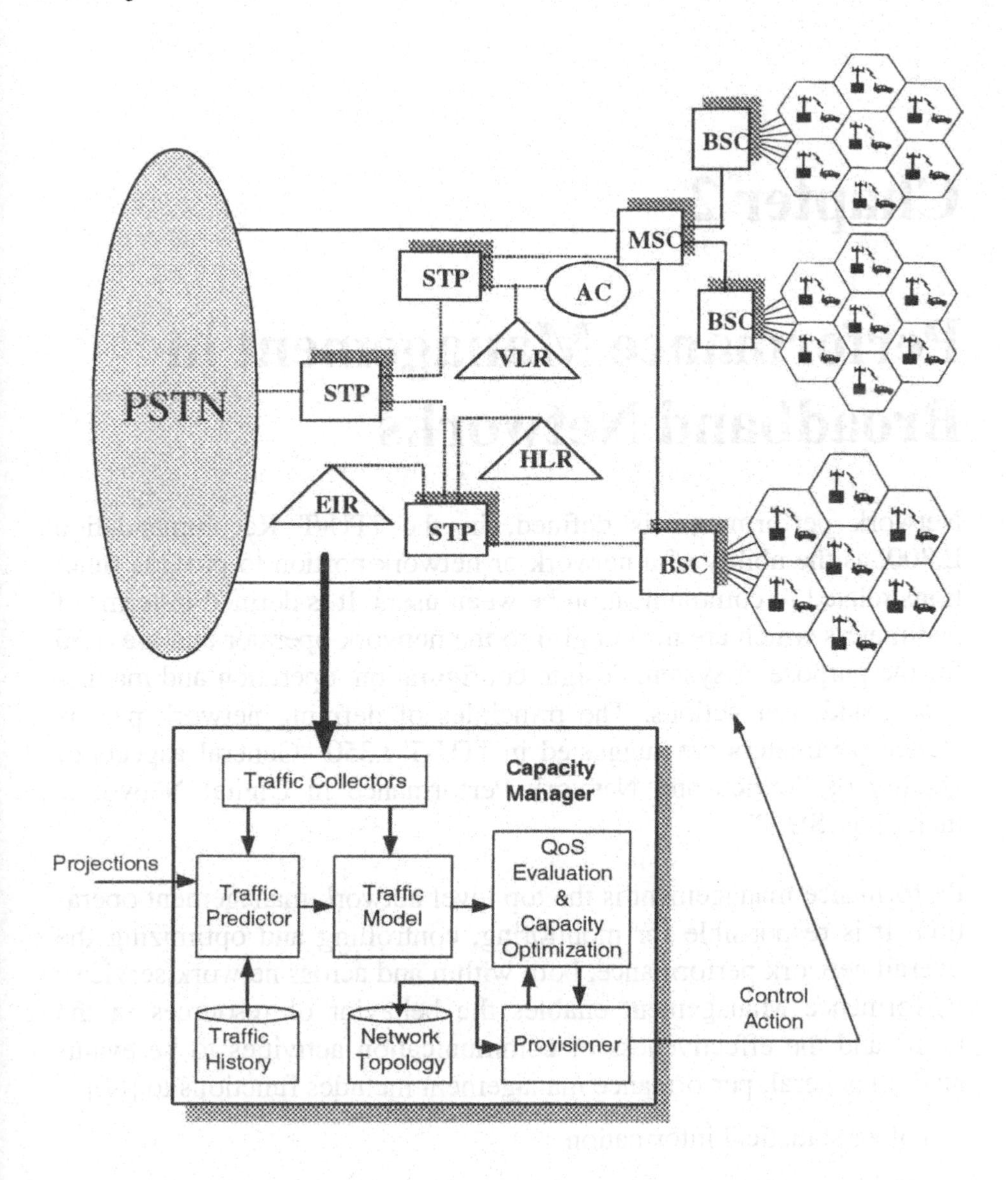

Figure. 1-2. The Proposed Framework.

Chapter 2

Performance Management in Broadband Networks

Network performance is defined, in the ITU-T Recommendation E.800, as the ability of a network or network portion to provide functions related to communication between users. It is defined in terms of parameters which are meaningful to the network operator and are used for the purpose of system design, configuration, operation and maintenance, and user actions. The principles of defining network performance parameters are suggested in ITU-T I.350 "General aspects of Quality of Service and Network Performance in Digital Networks, including ISDN".

Performance management is the top level network management operation. It is responsible for monitoring, controlling and optimizing the overall network performance, both within and across network services. Performance Management enables the behavior of resources in the OSIE and the effectiveness of communication activities to be evaluated. In general, performance management includes functions to [60]:

- gather statistical information
- maintain and examine logs of the system state histories
- determine system performance under natural and artificial conditions; and
- alter system modes of operation for the purpose of conducting performance management activities

Performance Management involves coordinating the actions of the lower level, task oriented applications such as fault, configuration, and traffic managers to recognize and resolve network performance problems. If we consider the traditional telecommunication management functions, i.e., operations, administration, maintenance, and provisioning (OAM&P), performance management encompasses the traffic management functions of operations, both network and traffic management functions of administration, and performance monitoring functions of maintenance.

There are two primary aspects of performance management: Monitoring (detection) and Control (resolution).

2.1 Performance Monitoring

Performance monitoring involves the continuous collection of data concerning the performance of the NE (and subsequently of the entire network). Very-low-rate or intermittent errors or congestion conditions in multiple equipment units may interact, resulting in poor service quality. Performance monitoring is designed to measure the overall quality of the monitored parameters to detect such deterioration (leading to a reactive form of control). It may also be designed to detect characteristic patterns before service quality has dropped below an acceptable level (leading to a preventative form of control). The basic function of performance monitoring is to track system, network, or service activities to gather the appropriate data for determining performance.

Performance monitoring in most of the standards is referred to as the monitoring of the transport resources. However some recent standards have generalized the term to include monitoring of switches, including traffic and QoS measures. **Transport performance monitoring** measures transmission error events detected by line, path, or section terminations. Examples of transmission error events that are monitored include counts over periodic time intervals of Cyclic Redundancy Check errors, frame slips, errored seconds and severely errored seconds. They are used as key elements in defining more severe impairment events or failure criteria, e.g., LOF (loss of frame). Associated with each of these error events is a parameter for measuring the performance. For example errored second ratio (ESR) is a performance parameter which is defined as the ratio of ES to total seconds in available time during a fixed measurement interval. Similarly severely errored second ratio (SESR) is another parameter. These parameters are used to evaluate the performance objectives of the paths. Table 1 of [62] specifies an example of the end-to-end objectives in terms of these parameters. The path fails to meet the error performance requirement if any of these objectives is not met.These parameters are further discussed in this chapter.

The **switch performance monitoring** deals with another part of network (switching instead of interoffice or loops) and at another level (traffic rather than physical characteristic). Traffic load is different from how traffic is physically carried. So performance characteristics that deal with traffic are quite different from the performance characteristics that deal with physical path transmission. Call blocking and delay are some of the parameters that are associated with the switch performance. Section 2.8 further elaborates on these performance parameters and associated performance objectives.

2.1.1 Performance Monitoring Functions

ITU-T M.3400 subdivides the functions of performance monitoring into generic functions, control functions and analysis functions [58]. The functions that deal with the monitoring of performance events and exceptions are listed under generic functions. Activities related to scheduling and modifying threshold data entities are related to the control functions. Functions that have the term correlate and analyze in their names relate to the analysis functions.

2.1.2 Role of Performance Monitoring

Performance Monitoring is used to monitor statistics concerning the health of the equipment and verifying the quality of transmission over a longer period of time.

- Performance Monitoring is closely related to alarm surveillance. Both are concerned with the detection of problems in equipment and transmission media. While alarms are designed to detect failure events, performance monitoring exception events indicate the rate of errors measured over a time interval that has exceeded a threshold.

- Performance monitoring is also related to fault localization and can be used for locating intermittent causes of degradation and anticipating failures that are preceded by gradual increases in error rates. When the statistical performance of a circuit drops from an acceptable level to an unacceptable, or significantly toward an unacceptable level, repair can be started, even if a customer has not reported the condition. Performance monitoring thus facilitates proactive servicing.
- It is also used in testing. Performance monitoring is used to verify the quality of newly installed or newly repaired facilities.
- In situations where a customer is guaranteed a level of performance for a transport service, performance exceptions can be used for billing rebates or verification of contractual requirements.

2.1.3 Performance Monitoring Requirements

All the above mentioned tasks demand certain capabilities and features in the NEs. A set of functional requirements has been defined in [59] for NEs in SDH based networks to carry out the performance monitoring tasks. These requirements are generic enough to include any digital transmission network and are therefore listed here for information purposes.

Performance data collection

Performance data collection refers to the event count associated with each of the performance parameters by the NE.

Performance monitoring history

Performance history data are necessary to assess the recent performance of transmission system. Such information can be used to sectionalize faults and locate sources of intermittent errors. Historical data, in the form of performance event counts, shall be stored in the NE.

Use of thresholds

The general strategy for the use of performance monitoring information and thresholds is described in recommendations [74],[75],[76].

This functionality in general involves:

- Threshold setting: The thresholds must be set in each NE via the OS (operations system [58]). The OS shall be able to retrieve and change the settings of 15-minute and 24 hour threshold values.
- Threshold crossing notification: As the threshold is reached or crossed for a given performance event, a threshold notification is generated. The detailed function is explained in [76].

Performance Data reporting

Performance data stored in the NE may be collected by the OS for analysis.

- Access by the OS to the performance data: Performance data shall be reportable across the OS/NE interface on demand when requested by the OS.
- Periodic report of performance data: Data collection may be performed periodically to support trend analysis to predict future failure or degraded conditions. On request of the OS, performance data of specific ports shall be reported periodically.
- Autonomous report on reaching or crossing a threshold: Performance data shall be reported across the NE/OS interface automatically upon reaching or crossing a threshold.

Performance monitoring during unavailable time

Performance event count shall be inhibited during unavailable time as defined in [62].

2.2 Performance Control

Performance control involves the administration of information that guides or supports the network management function and application or modification of traffic controls to help provide network management. As mentioned earlier, control could be of reactive or preventative form. Performance control actions generally consist of issuing commands which either utilize spare network capacity (e.g. alter communication paths) or confine the effects of the problems (e.g. limit network traffic). Examples include downloading switch controls, issuing cross connect commands, switching to backup equipment and exercising admission control.

Some of the control aspects of performance management are provided in the later part of this book.

2.3 Performance Monitoring in T-Carrier Systems

The T-carrier system was the first digital system introduced into the public networks. A carrier system, in general, is characterized by a transmission component, a (user) interface component, and a user termination equipment component. The transmission component is a transmission medium carrying multiple channels. T-carrier systems use twisted pair copper facilities as a transport medium and carry 24, 96, or 672 simultaneous channels operating at 64-kbps each (the T1, T2, and T3 systems, respectively). It should, however, be noted that the digital systems are now transmitted using a multitude of media, including paired cable, coaxial cable, radio, satellite, and optical fibre. The nomenclature, thus, for describing generic digital carrier systems, regardless of underlying medium, is DS1, DS2, and DS3. These generic digital hierarchies are called PDH (Plesiochronous Digital Hierarchy) to distinguish them from the nascent SDH (Synchronous Digital Hierarchy).

This section introduces the performance management aspects and issues of PDH, in general, and the T-carrier systems, in particular. The section starts off with highlighting the main characteristics of T-carrier systems. Thereafter, the format of the signal transmitted over the facilities is presented and all the performance related error events are discussed. Finally, an exhaustive list of all the anomalies, defects, events and parameters associated with the performance of the T-carrier systems is provided.

2.3.1 T-Carrier System: an Overview

2.3.1.1 T1 Fundamentals

T1 is characterized by the following [86]:

- A four wire circuit: two wires for transmit and two for receive
- Full Duplex
- Pulse Code Modulation
- Time Division Multiplexing
- Framed Format: As pulse code modulation scheme is used, the 24 channels are time-division multiplexed into a frame to be carried along the line. Each frame represents an 8-bit sample from each of the 24-channels.
- Bipolar Format: Every other pulse is represented by the negative equivalent of the pulse.
- Byte Synchronous Transmission: Timing for the channels is derived from the pulses that appear within the samples. This timing keeps everything in sequence. Should the devices on both ends of the line not see any pulses, they would lose track of where they were and **slippage** would occur. When dealing with a T1, the bit sequencing is easier to derive. However, at multiplexing schemes above the T1 rate, additional bits are inserted in the data stream to maintain a constant clocking reference. The use of these overhead bits (**bit stuffing**) brings both transmitter and receiver up to a common signalling speed to maintain frame synchronization.
- Channelized or Non-Channelized: Generally, T1 is 24 channels of 64 kbps each plus 8 kbps of overhead. However, bursty data is also possible.

2.3.1.1.1 Extended Super Frame (ESF) Format

T1 uses some very specific conventions to transmit information between both ends. One of these is a framing sequence that formats the samples of voice or data transmission. It is easy to think of a frame in terms of the 8-bit samples of the 24 channels being strung together in a logical sequence. After the 192 bits of information are compiled, a framing bit is added, creating a frame of 193 bits of information. The framing bit can be equated to a pointer or address. Since the line is moving information at 1,544,000 bits per second, it would be very easy to skew left or right and deliver the information out of sequence. Therefore, the extra 8000 bits of information per second create a locator on which the equipment can lock in. This pointer allows the devices to read a pattern of bits to know which frame is being received or transmitted and the location of each channel thereafter.

Framing has undergone several evolutions over the years. Starting from D1 frame there have been D2, D1D, D3, D4, D5 and ESF (Extended Superframe). Among these, ESF standard is the most common and widely accepted one. ESF facilitates availability, error checking, real time diagnostics, and performance monitoring. The 8-kbps of overhead originally used strictly for framing is now subdivided into three functions. Only 6 of the 24 repeating framing bits are used for framing synchronization. Six additional bits are used for error detection by employing a cyclic redundancy check (CRC6). This leaves 12 of the framing bits for a facility data link communications channel (4-kbps). Using this shared resource, the carrier has the option to perform the three functions necessary to support customer needs. These are:

- Framing Synchronization
- Error detection
- In-service monitoring and diagnostics

2.3.1.1.2 Framing

Framing is important since its loss can affect performance by the incorrect synchronization (channel 1 could wind up connected to channel x [2-24] if framing is incorrect). Errors may cause a loss of frame (LOF) that would disrupt the circuit. Far-end or intermediate equipment could be thrown into a loss-of-frame condition and cause lost data. A loss of frame requires equipment to re-frame. While the devices along the circuit are re-framing, data throughput is disrupted for a period of time.

2.3.1.1.3 Cyclic Redundancy Check-CRC6

Each block or frame in a digital transmission system is monitored by means of an inherent Error Detection Code (EDC), e.g. Bit Interleaved Parity (BIP e.g. BIP-8) or Cyclic Redundancy Check (e.g. CRC-6,

CRC-4). T1 does CRC6 which is designed to detect errors that occur on the line. Using CRC6, an entire ESF (24 frames of information or 4632 bits) is checked for accuracy. This check is a number created by the CSU (customer service unit) by performing a mathematical computation on the 4632 data bits. The result is transmitted to the distant end in the next frame. The receiving CSU calculates the result using the same mathematical operations and if the results are different an error has occurred. The use of CRC6 allows an error detection with 98.4% certainty.

Errors are classified in various categories based on performance as outlined in the next section. In general ESF error events are processed to derive

- Errored seconds (ES). At least one ESF error exists in a 1-second period.
- Severely errored seconds (SES). A 1-second period where 320 or more ESF errors have occurred.
- Failed seconds or failed signal state (FSS) occurs when 10 consecutive SES's occur and clears when 10 seconds of data are processed without an SES. A failed second is counted whenever an FSS exists. Information on ESF errors is stored in buffers within the CSU for later retrieval.

By monitoring SES events for both directions at a single path end point, a network provider is able to determine the unavailable state of the path. According to [62], a path is unavailable at the onset of 10 consecutive SES events in both directions.

2.3.1.1.4 The Facility Data Link

The 4-kbps overhead set aside for facility data link control is a synchronous communications channel. This channel is mostly used for the exchange of information between equipment devices along the circuit. The carrier or end user can communicate with the remote equipment (CSU) using maintenance message format. Some of the maintenance messages allowed, include [86]:

- Send one hour performance data reports: This will include the existing status of the link, how long it has been since the last check (up to 24 hour), errored seconds and failed seconds in the present 15 minute cycle, overall 24-hour cycle, and the past four 15-minute cycle performances.
- Send 24-hour ES performance data: The CSU will send specific error events logged within the past 24 hours, in 15 minutes increments.
- Reset registers: Empty the buffers and start counting errors all over again.
- Send errored ESF: The CSU sends the current count of errors accumulated (up to a maximum of 65,535).
- Reset ESF register: Empty the buffer and start at 0.

2.3.2 T3 Fundamentals

- Full Duplex
- Bit Stream or Bandwidth of 44.736 Mbps: This is the multiplexed stream of twenty-eight T1s into a single digital carrier, plus the 1.5-Mbps overhead for channelization, frame identification, and more.
- Supports Channelized as well Non-Channelized service. Non-channelized service is desirable for high speed data communication, full-motion video conferencing etc.
- Pulse Code Modulation
- Time Division Multiplexing
- Bipolar Signal with return to zero
- Carrier only service
- Framed Format. T3 uses a 4760-bit frame which is much larger than 193-bit T1 frame. The T3 frame is known as M13 frame, which comes from the use of M13 multiplexer. The T3 standard rules (protocols) deal with the M13 framed format transmitted in an asynchronous mode. Due to certain problems with this mode of transmission,

three newer methods are in various stages of deployment which are C-bit parity, Synchronous Transmission (SYNTRAN), and Synchronous Optical network (SONET)

2.3.2.1 M13 Frame

The framed format for a T3 uses the M13 frame. When first employed in the network, the M13 framed format used an asynchronous protocol. Asynchronous implies that the timing for the digital bit stream is ambiguous i.e. there is no synchronization on the transmission of the bit stream. Therefore, the transmitter and receiver must provide bit stuffing (resulting in extensive pulse stuffing overhead) to come up to a common clocking arrangement. The M13 multiplexing is performed in two steps

- Four DS1 signals are multiplexed together using pulse stuffing synchronization to form an internal DS2 signal.
- Seven DS2 signals are multiplexed together using a fixed pulse-stuffing synchronization to form a DS3 signal.

The DS3 frame is composed of 4760 bits of which 4704 are for user information and 56 are for overhead.

- Twenty-eight bits are reserved for control (called F-bits).
- Two bits are used for low-speed signals between ends (called X-bits).
- Two bits are used for parity checks on the user information. The transmitter counts the user data stream and inserts a 2-bit parity sequence (P-bits)
- The use of multiple frames in the DS3 signal requires multi-frame alignment similar to framing bit sequence (M-bits = 3).
- There are 21 C-bits in the frame that are used depending on the protocol being used.

The M13 asynchronous protocol uses C-bits as stuffing bits in 7 of 18 repeating frames to get common timing and clocking for seven DS2s that are multiplexed together into a DS3. Thus, there is very little channel capacity left for signaling and maintenance using this M13 asynchronous frame structure. To overcome these deficiencies some new methods have been proposed which involve using C-bits for purposes other than to indicate if stuffing has or has not occurred. Some of these new methods include C-Bit parity operation and SYNTRAN. Under C-Bit parity operation, a concatenated state of three C-bits (e.g. 7, 8, and 9) is used that yields a 28.2-kbps channel capacity. This feature enhances the signaling and maintenance capabilities. SYNTRAN, on the other hand, uses a super frame format and concatenates the M13 4,760-bit frame into 699 frames per super frame. Using this protocol, the C-bits are redefined to support greater diagnostics, non-disruptive monitoring, and the isolation of sub-rated multiplexed signals inside the DS3.

2.3.3 Performance Monitoring

This section lists the anomalies, defects and parameters detectable with T1 (DS1) & T3 (DS3) transmission system.

2.3.3.1 Performance Monitoring Data

In-service anomaly conditions are used to determine the error performance of a PDH path when the path is not in a defect state. The following two categories of anomalies are defined:

- An errored frame alignment signal
- An EB (Errored Block) as indicated by an EDC (CRC6)

In service defect conditions are used to determine the change of performance state which may occur on a path. The three following categories of defects related to the incoming signals are defined:

- Loss of signal (LOS)
- Alarm indication signal (AIS)
- Loss of frame alignment (LOF)

DS1 Performance Parameters

- Bit Error Rate for path (BERPI)
- Coding Violations Count for path (CVPI)
- Cyclic Redundancy Error Check (CRC) Error Count

- Burst Errored Seconds Count for Path
- Frame Format Status
- Frame Bit Error Count (FBE)
- Errored Seconds Count for path (ES-P)
- Severely Errored Seconds Count for path (SESPI)
- Severely Errored Framing Seconds Count for Path
- Unavailable Seconds Count for path (UASP)
- Estimated Slip Count for Path (SLIPC)

DS3 Performance Parameters

- Bit Error Rate for line (BERLI), for path (BERPI)
- Coding Violations Count for line (CVLI), for path (CVPI)
- Frame Format Status
- Frame Bit Error Count (FBE)
- Frame Parity Error Count
- Far End Block Error Count (FEBE)
- Errored Seconds Count for line (ES-L), for path (ES-P)
- Severely Errored Seconds Count for line (SESLI), for path (SESPI)
- Unavailable Seconds Count for path (UASP)
- Monitored Seconds Count Line (MSL)

The available remote in-service indications such as RAI (Remote Alarm Indication) or, if provided, FERF and FEBE are used at the near-end to estimate the number of SES occurring at the far end.

2.4 Performance Monitoring in SDH/SONET based Networks

SDH and SONET are the two standards associated with the Synchronous Digital Transmission. Synchronous Digital Transmission is a dominant technology and when used with network elements such as

Add/Drop Multiplexer (ADM) and Digital Cross-Connect (DXC) make possible the formation of more effective point to multiple point network configurations. The introduction of SDH/SONET along with ring topology has added various new features to networks. Other features of this technology include survivable SONET network architectures that incorporate self healing capabilities. These architectures include point-to-point systems with diverse routing and Automatic Protection Switching (APS), SONET self-healing rings, and SONET mesh-type self-healing network architectures based on re-configuration of Digital Cross-Connect systems. These advanced architectures ensure that the network management resources are allocated effectively and conflicting resolution paths are not pursued.

This section is devoted to analyzing the performance management aspects of the SDH/SONET based telecommunication networks. The section starts off with a brief overview of the SDH and SONET signal formats. From a network management viewpoint, SDH/SONET contains some features that are quite useful in providing visibility of the level of performance of the telecommunication system. One of the key features is the layered architecture of the synchronous format defined in SDH/SONET specifications. This layered architecture can be potentially exploited to develop new ways of managing future networks. This section, therefore, also provides some understanding of this layered architecture and points out some, not so obvious, built in management features. Finally, an exhaustive list of all the parameters associated with the performance of the SDH/SONET based transmission is provided.

2.4.1 SDH/SONET: an Overview

SDH, i.e. Synchronous Digital Hierarchy, [59], [61] is a digital transport structure that operates by appropriately managing the payloads and transporting them through (synchronous) transmission networks. SDH is composed of STM-n signals where n is a fixed number and indicates the number of multiples of the base STM-1 that has a bit rate of 155.520 Mbps. The term synchronous in SDH comes from the fact that the process of multiplexing plesiochronous tributaries (i.e. conventional PDH signals such as DS-1, DS-2, DS-3 and DS-4) into STM-n adapts a synchronous multiplexing structure. SONET [4] is the new standard family of transmission interfaces for telephone company optical networks. It provides synchronous transmission from a basic rate of 51.84 Mb/s (OC-1) up to 2.49 Gb/s (OC-48) in increasing multiples of the basic rate.

SDH and SONET maintain an extremely close relationship and the description of SDH inevitably includes a treatment of SONET. However minute differences do exist such as the base rate of SONET is 1/3rd the base rate of SDH. They also differ in terms of diversity of transmission rates, frame format, intermediate signal units, and multiplexing structures. These differences are not critical from performance management perspectives and, therefore, this discussion refers mostly to SDH, assuming that the SONET issues are addressed implicitly.

2.4.1.1 Layered Overhead Structure

The transmission handling procedures of SDH/SONET transmission are systematized in accordance with the layering concept [84], [91], and this is reflected in the formats of overheads. From a network management viewpoint, the importance of the layered structure lies in the types of information provided at each layer and the ability to use that information in determining network performance.

The four basic layers of this layered architecture and how they relate to equipment and specifications are as follows:

- Physical/Photonic layer - defines the optical pulse shape, power levels, and wavelength
- Section layer - provides signal framing and basic-level performance monitoring of the payload
- Line layer - provides protection switching and multiplexing functions for the information payload
- Path layer - provides signal labeling and tracing for end-to-end payload management

Each of these layers, except the photonic layer, contributes to or processes an associated overhead in the transmitted signal frame to support the signal processing or network management functions. Figure 2-1 depicts an STS-1 format, with the path layer having two overhead functions; the Synchronous Transport Signal (STS) path overhead and the Virtual Tributary (VT) path overhead.

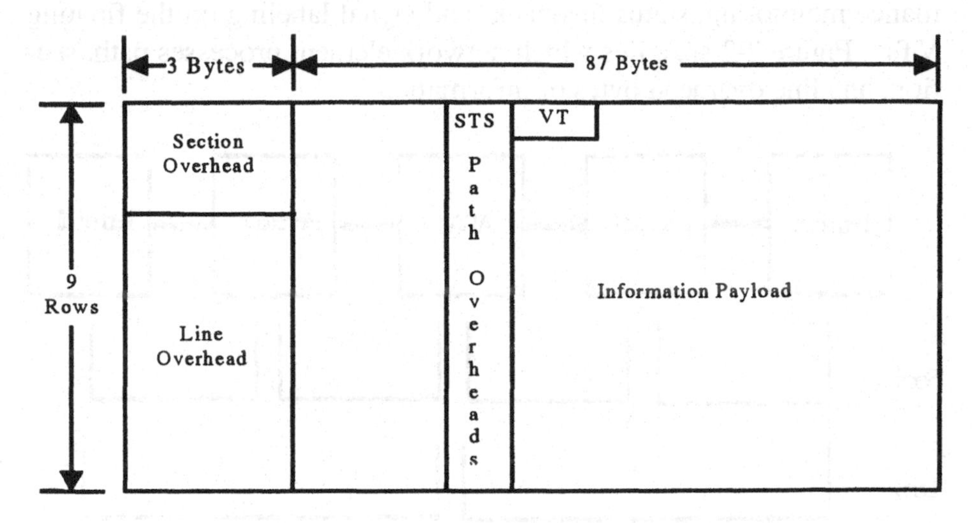

Figure. 2-1. STS-1 frame format and overhead layers

The three overhead layers contain information that can be used to manage the network in a hierarchical manner. Each network element that performs Path Terminating Element (PTE), Line Terminating Element (LTE), and Section Terminating Element (STE) functions on the payload must be capable of processing the overhead information for signal processing and network management support associated with that layer.

Section overhead must be processed by each network element to accomplish basic transport function of framing on STS-1 signals. In addition to signal framing information, section overhead also contains the basic level performance data. Functions supported by the line layer for payload management consist of STS-1 payload pointer storage and automatic protection switching commands to be processed by line and path terminating elements. Network management functions supported at the line layer consist of performance monitoring data, an express orderwire channel and a 576kb/s DCC (Data Communication Channel). Path overheads are used to provide end-to-end management of payloads at the service terminating location of the SONET networks.STS-1 path overhead provides performance monitoring, status feedback, and signal labelling, user channel and STS tracing function. VT path overhead, which is provided with a floating format of VTs provides performance monitoring, status feedback, and signal labeling on the floating VTs. Figure 2-2 specifies which network element processes path, section, and line overhead bytes of information.

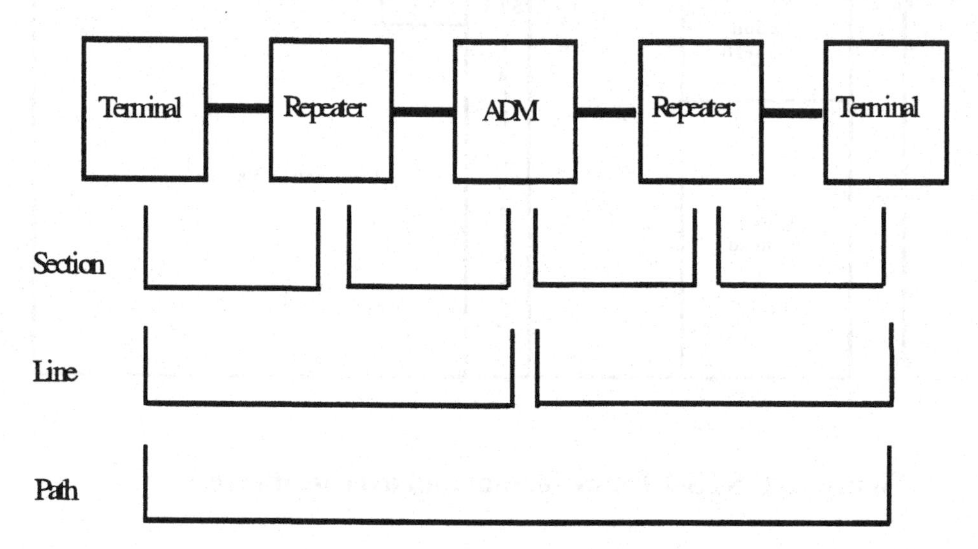

Figure. 2-2. Section, Line, and Path overheads with network elements

2.4.2 Performance Monitoring

2.4.2.1 Performance Monitoring Locations

In the SDH management context, monitored points coincide with termination points. Several monitoring points have been identified for SDH performance monitoring purposes [59], [61] and they include:

- Regenerator section termination point
- Electrical and Optical physical interfaces termination point
- Multiplex section termination point
- Protection switch

- Path termination
- Multiplex section adaptation

2.4.2.2 Performance Monitoring Data

Until recently, performance data could consist of any information available from a NE used to indicate how well it performed its functions i.e. alarms and statuses. The performance monitoring primitives, parameters and failure criteria for DS1, DS3, and SDH/SONET network elements, however, now also have been described and standardized [59], [61], 62], [64]. This section will briefly discuss the primitives and parameters available from basic information provided in the SDH/ SONET overhead architecture.

2.4.2.3 SDH/SONET Performance Primitives

The performance primitives are the basics used at each overhead layer in building parameters to provide management information. The following list of these primitives is divided into anomalies, defects and failures [91].

The performance anomalies are:

- Section Framing word Error (FE) - framing word error in the framing pattern
- Bit Interleaved Parity (BIP-N) - parity error code generated for comparison at receiver to determine transmission integrity
- Path Far-End Performance Report (FEPR) - path status message sent from receiver to transmitter

The performance defects are:

- Section Severely Errored Frame (SEF) - four consecutive errored frame alignment signals followed by two successive error-free frames
- Loss Of Signal (LOS) - detection of all “zeroes” pattern on the incoming OC-N signal
- Loss of Frame (LOF) - detection of an SEF for a 3 ms period

- Loss of Pointer (LOP) - detection of an invalid STS or VT pointer
- Line Alarm Indication Signal (Line AIS) - section terminating equipment notifying line terminating equipment of a failure upstream
- STS Path Alarm Indication Signal (STS Path AIS) - line terminating equipment notifying path terminating equipment of a failure upstream
- Virtual Tributary Path Alarm Indication Signal (VT Path AIS) - VT Path terminating equipment notifying VT path terminating equipment of a failure upstream
- Line Far-End Receive Failure (FERF) - generation of a code by far-end line terminating equipment to notify signal originator of AIS received

The Performance Failures are:

- Loss of Frame Failure - declared when 2 to 10 consecutive 1-s intervals contain an SEF event
- Loss of Signal Failure - declared when LOS state persists for 10 to 20 consecutive 1-s intervals
- Loss of Pointer Failure - declared when 2 to 10 consecutive seconds contain LOP defects
- AIS Failure - declared when 2 to 10 consecutive 1-s intervals have any AIS present
- Line FERF Failure - declared when incoming line FERF last for 2 to 10 consecutive seconds
- STS-Path Remote Alarm Indication Failure (RAI) - declared within 5 ms of detecting incoming STS-Path RAI signal
- VT-Path Remote Alarm Indication Failure (RAI) - declared within 5 ms of detecting incoming VT-Path RAI signal

2.4.2.4 SDH/SONET Performance Parameters

Performance parameters are derived by processing primitives and are essentially counts of various impairment events accumulated over a specified collection period (i.e. 15-minute interval etc.).

The following list briefly describes the parameters [91]:

- Coding Violations for Section (CV-S), Line (CV-L), STS Path (CV-P), and VT Path (CV-V) - Bit Interleaved Parity (BIP-N) errors detected on incoming signal at each layer
- Errored Seconds for Section (ES-S), Line (ES-L), STS Path (ES-P), and VT path (ES-V) - a second during which at least one coding violation has occurred at that layer
- Severely Errored Seconds for Section (SES-S), Line (SES-L), STS Path (SES-P), and VT Path (SES-V) - a second with a specified number of coding violations where the specified values are set in relation to the layer having errors
- Severely Errored Framing Seconds For Section (SEFS-S) - a count of 1-s intervals containing one or more SEF events
- Unavailable Seconds for Line (UAS-L), STS Path (UAS-P), and VT Path (UAS-V) - count of 1-s intervals where the associated layer is not available
- Loss of Signal Seconds (LOSS) - count of 1-s intervals containing one or more LOS defects
- Alarm Indication Signal Seconds (AISS) - count of 1-s intervals containing one or more AIS defects
- Loss of Pointer Seconds (LOPS) - count of seconds during LOP detection
- Far-End STS-Path and VT-Path Layer Parameter - performance of the far-end path is conveyed back to the near end via summarized status bits in the overhead
- Protection Switch event (PS) - declared when a protection system switches from a working line to protection line

- Protection Switching Counts (PSC) - a count, for each line, of protection switching events
- Protection Switching Duration (PSD) - a count of seconds service was removed from line being monitored
- Consecutive SES counts configurable in the range of 2 to 9 SES (CSES)
- Administrative Unit Pointer Justification Event (AU PJE)

A general description of these monitored events and their threshold values are specified in [62].

2.4.2.5 SONET Maintenance Signals

Standardization of maintenance signals requires line, section, and path terminating elements to make decisions on conditions of the received payloads. Specific actions taken by the network elements should be reported as status information to the management system. Information available at the line and path layers consists of [91]:

- STS and VT Synchronous Payload Envelope Unequipped Indications - provides status reporting on partially equipped network elements
- Line, STS Path, and VT Path Alarm Indication Signal (AIS) - indicates loss of signal condition on upstream network elements
- Far-End Receive Failure (FERF) - a message returned to transmitting network elements of AIS or incoming line failure being received
- Yellow Signal - indicates receipt of AIS to upstream network elements of the same layer peer level

2.5 Performance Monitoring in ATM Networks

The digitization of most of the public telephone networks worldwide, as well as the concurrent deployment of optical fiber, now allow much wider bandwidths (higher bit rates) to be used than previously has been the case for these voice and data networks. With bit rates of 155 Mbps, 622 Mbps, and 2.4 Gbps now available, it is natural to start planning new services that would make use of these bandwidths. Prime candidates include applications involving images, video and multimedia. In addition, it becomes natural to start merging services provided by different networks (e.g. voice and packet data) onto one common network. Broadband Integrated Services Digital Networks (B-ISDN) have been developed that are capable of carrying out the functions just described.

ITU-T has issued a number of recommendations concerning B-ISDN. These recommendations include adoption of ATM (Asynchronous Transfer Mode) as the transport mechanism for the B-ISDN. Actively involved in standards development is, obviously, the ATM forum, a voluntary group of manufacturers, vendors, communication carriers, and other organizations with interests in seeing ATM standards development speeded up to expedite the delivery of ATM products. These players have also been engaged in a focused effort to specify and standardize the operations functions that would allow broadband network providers to effectively and efficiently manage their high-speed, high-performance ATM networks. The main motivating factor behind this effort is the general understanding that even high-performance ATM-based networks will, at times, experience failures, congestion, and periods of degraded performance. Thus, without the capabilities to address the performance awareness by detecting all performance degradation within a network before the services are actually affected, and subsequently overcome these problems, the viability of the broadband networks will be in jeopardy.

This section summarizes the current state of the performance management being addressed in the recommendations for ATM networks by ITU-T, T1 committee, and ATM forum. The section starts off with a brief introduction of the B-ISDN protocol reference model and then addresses the performance monitoring aspects of the ATM networks.

2.5.1 ATM: an Overview

ATM has been designated as the target transfer mode approach to provide the desired integration of the various traffic types to be supported by B-ISDN. It is essentially a packet-switched mode of transfer through the network, using short, fixed length, 53 octet (byte) packets called cells. The cells are assigned on demand at the user-network interface (UNI). B-ISDN rates have initially been chosen as 155.52 Mbps and 622 Mbps (two of the SONET rates), with the possibility of eventually reaching 2.4 Gbps and above. The ITU-T and other standard bodies have identified five generic service classes that an ATM B-ISDN might support. These include CBR, VBR-rt (Variable Bit Rate - Real-Time), VBR-nrt (Variable Bit Rate - Non Real-Time), ABR (Available Bit Rate), GFR (Generic Frame Rate) etc.

To provide these above-mentioned classes of services, a layered protocol model has been developed for B-ISDN [65] and is shown in Figure 2-3. The physical layer, in this model, is medium dependent. For example, SONET would presumably provide the physical layer capability for optical fibre in North America. The ATM layer is responsible for generating the asynchronous 53-octet cells. It is also responsible for cell routing, (de)multiplexing user traffic over a link or path, verifying cell header integrity, and limited flow control capability. Since the B-ISDN traffic classes arise from very different sources, traffic from each class must be handled differently in preparation for generating its corresponding ATM cells. For this reason an ATM adaptation layer (AAL) has been defined to generate 48-octet segments from user information passed down from the higher layers. These segments, in turn, have five header octets added at the ATM layer to form 53-octet ATM cells transmitted out over the network using the physical layer.

Two types of ATM cells have been defined, as shown in Figure 2-3 [66]. One type corresponds to cells generated at either side of user-network interface (UNI) i.e. the access to the network. The other type of cell represents cells switched at a network node interface (NNI). Most of the 5-octet header, in these ATM cells, is devoted to providing a routing or connection id (ATM is designed primarily as a virtual circuit-type, or connection oriented service). End-to-end connections, or virtual circuits, are defined in terms of virtual path id (VPI) and a virtual connection id (VCI). A virtual path (VP) represents a network-defined, end-to-end connection that follows a specific route or path and provides a specified quality of service (QoS). The QoS could include such parameters as bandwidth available over that path (this may be minimum, maximum, average, or peak bandwidth), constraints on time delay and time jitter, and cell loss probability etc. Each VP has multiple VCs within it. VCs are typically per service oriented, and terminate on Customer Equipment. Cells, thus, are routed on the basis of VPI/VCI pair. The other fields of these ATM cells include a three-bit payload type (PT) that is used to designate the type of information, i.e. user or operations/management (OAM) data, being carried in the 48-octet cell information field. It is also used to indicate whether or not congestion has been experienced. The 8-bit HEC is used to provide the error checking of the 5-bit header. The CLP bit, if set to 1, indicates a low priority cell that can be dropped in the event of network congestion (refer to [69] for details of traffic and congestion control functionality).

2.5.2 Operations, Administrations, and Maintenance (OAM) Flows

As mentioned in the previous section, the ATM cells are used for carrying either user information or operations and management (OAM) data. The reason for generating these OAM cells is that the broadband operations sometimes require the exchange of operations and management information between various nodes in the network. Specific information such as failure indications, performance monitoring data, test requests, system protection, and fault localization will need to be communicated between peer (ATM and Physical) layers at various nodes in the network. Mechanisms to transmit such information are referred to in ITU-T standards as F1-F5 flows [67] and are performed on five OAM hierarchical levels associated with the ATM and physical layer of the ITU-T reference model. These flows are made possible by specially marked ATM cells that contain the relevant information and are referred to as operations, administration, and maintenance (OAM) cells. As mentioned earlier, these OAM cells are distinguished from user-data cells by payload-type (PT) indicator in the header.

The format of F4 (VP level) and F5 (VC level) OAM cells is shown in Figure 2-3. In these OAM cells, the OAM Cell Type (4 bits) indicates if the cell is being used for performance, fault, or configuration management. OAM Function Type (4 bits) identifies a specific performance, fault, or configuration management function. For example, in performance management, this field identifies if an OAM cell contains forward monitoring or backward monitoring of performance data, or both. Cyclical Redundancy Check (10 bits) is the error detection code (CRC-10) for the OAM cell payload.

Each VPC and VCC can have an operations channel (bandwidth allocated) through which these OAM cells can be transferred. The performance information of the ATM network can thus be monitored continuously and transferred by OAM monitoring cells periodically. Generally two types of operations flow would exist in an ATM-based network.

- End-to-End operations flows refer to OAM cells that are used to communicate operations information across an entire end-to-end VPC (or VCC). Such OAM cells may be inserted and monitored at intermediate nodes along the end-to-end connection, but can be terminated (processed and extracted) only at the connection endpoints.
- Segment operations flows refer to the OAM cells that are used to communicate operations information within the bounds of a single VP/VC link or group of interconnected links, all of which are under the control of a single administration. As in the case of end-to-end operations flows, intermediate nodes can monitor or insert cells but the termination of cells can only take place at the segment end points.

2.5.3 Performance Monitoring

Although ATM-based networks are high-performance networks, performance degradation may still occur at times due to reasons such as transmission errors or intermittent troubles experienced by an underlying physical facility; faulty protocol implementations and software problems at the ATM layer; and temporary actions taken by ATM switches to alleviate congestion. To facilitate early observation of slowly degrading VPC and VCC performance, a performance monitoring scheme is agreed to in standards [68], [90], [88], [89], 62]. Each layer in the ITU-T B-ISDN protocol reference model of Figure 2-3 is assigned functionality to support performance monitoring. The following subsections describe these standardized performance monitoring aspects of ATM networks.

2.5.3.1 Monitoring Locations

The nodes that constitute an ATM virtual circuit are the customer premises equipment (CPE), virtual channel (VC) switches, and virtual path (VP) Cross-connects. These nodes are defined by the underlying physical layer. Any or all of these nodes could be monitored for performance. In literature [90], the performance monitoring locations are generalized as follows:

- Near-End Monitoring provides the performance of a received signal from its origination to its termination. The monitoring point is at the received signal termination.
- Far-End Monitoring provides the performance of a transmitted signal from its origination to its termination; performance at the far-end termination is sent back to the monitoring point in received signal overhead. The monitoring point is at the received signal termination where the overhead terminates, or at an intermediate point where the overhead is read.
- Intermediate Monitoring is at intermediate locations in a transport node such that near and far end performance indicators are read but not terminated.

It is quite clear that the support for far-end monitoring in B-ISDN networks facilitates the detection of trouble and its direction on the virtual circuit, and by comparing intermediate monitoring results for each direction, the side of a NNI or UNI that is the source of trouble can also be determined.

2.5.3.2 Performance Monitoring at TC sublayer

As discussed earlier in this section, ATM cells are mapped into the physical layer (SONET/SDH) payload by the Transmission Convergence (TC) sublayer of ATM adaptation layer (AAL). The TC sublayer performs monitoring at the physical layer path terminations where received cell headers are processed, and aggregates it across all virtual links in the physical layer path.

All the performance indicators, parameters, and alarms associated with this TC sublayer are listed below [90].

Indicators

- Cell Delineation Mechanism: TC sublayer uses the ATM cell Header Error Check (HEC), and based on the number of errors in that field signifies an Out-of-Cell Delineation (OCD) condition. OCD implies loss of contiguous cells. If OCD condition persists for a few seconds, a Loss-Of-Cell Delineation (LCD) alarm results.
- Cell Header Integrity: TC sublayer uses Header Error Check (HEC), VPI/VCI validity check, and Cell Loss Priority (CLP) to detect and discard cells with invalid headers or low cell loss priority (CLP set to 0).

Parameters

- Cell Loss rate: Cells discarded due to header errors or congestion
- OCD seconds: Seconds with one or more OCD event

Alarm

- Loss of Cell Delineation (LCD)

2.5.3.3 Performance Monitoring at ATM layer

For VPCs/VCCs, cell payloads are monitored over an entire connection or over segments of a connection, using OAM cells. VPC/VCC monitoring provides a measure of end-to-end QoS, while VPC/VCC segment monitoring from edge-point to edge-point in a specific administration provides trouble sectionalization. All the indicators, parameters and alarms associated with performance monitoring at this ATM layer level are listed below.

Indicators

- Continuity Check Mechanism: This check detects VPC/VCC specific trouble such as corruption of the VPI/VCI address tables needed in each ATM node to establish VPCs/VCCc. This check is activated on selected VPCs/VCCs, and consists of sending a continuity check OAM cell from an end-point for a certain duration, and if the other end does not receive any cells for that duration, this signifies an impending Loss of Continuity (LOC).
- OAM cell content: The near end of the connection generates a Bit Interleaved Parity (BIP-16) Error Detection Code (EDC) over a block (128, 256, 512, 0r 1024 cells in length) of user data cells. This EDC and block size is inserted in the payload of an OAM cell which is (optionally) time-stamped and transmitted immediately (forward monitoring) following the user data block. As the cells are received at the far end, EDC is recalculated and compared with the EDC contained in the received OAM cell. A mismatch would indicate that the block has experienced an error. In addition, the received block size will be compared with the block size specified in the incoming OAM cell to see if cells were mis-inserted or lost during transmission. The time-stamp of the incoming OAM can be used to compute the cell transfer delay or delay variation. These findings may be recorded locally or, again, using OAM cells, sent back to the originating end (backward monitoring).

Parameters

- Errored Cell Blocks Rate
- Severely Errored Cell Blocks Rate
- Cell Loss Rate
- Cell Mis-insertion Rate
- Cell Transfer Delay (CTD)
- Cell Delay Variation (CDV)

Alarms

- Loss of Continuity (LOC)

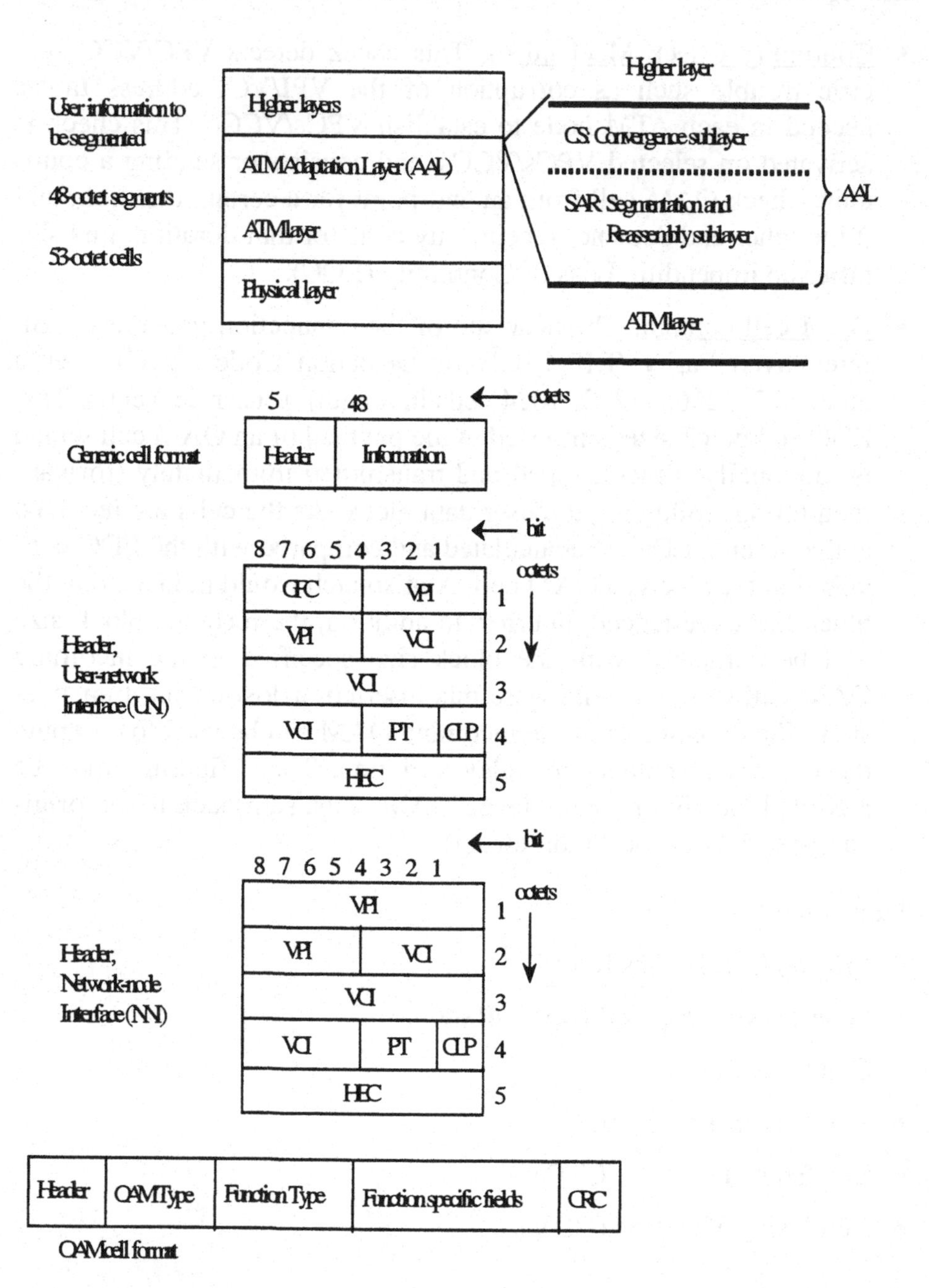

Figure. 2-3. B-ISDN Protocol Reference Model, and ATM Cell Headers

2.6 Performance Monitoring in Frame Relay Networks

Frame Relay is a connection-oriented, high-speed, packet-based technology that supports variable length packets and is considered as the next logical step of the X.25 standard. Frame Relay protocol supports data rates from 64 kbps to 2 Mbps. Just like X.25, it regulates the interface between the customer computers and the networks, and is implemented in products such as LAN bridges, routers, and T1 multiplexers. Frame Relay aims at reducing network delays, increasing throughput, providing more efficient bandwidth utilization, and decreasing communication equipment cost. All these benefits are achieved by minimizing formats and procedures. For example, in Frame Relay the network layer (layer 3) of X.25 has been completely eliminated, statistical multiplexing capability has been added to the data link layer (layer 2), and the functionality of the data link layer has been reduced by removing error correction and retransmission capabilities.

Some of the rival technologies of Frame Relay include SMDS (Switched Multimegabit Data Service), T1 links, and switched DS-1. Frame Relay, along with Cell Relay (such as ATM, which is the underlying technology of the B-ISDN), are the main foci of current high-speed communications. Some vendors treat Frame Relay and Cell Relay as competing technologies while others treat them as complementary and are merging the two.

Frame Relay can be obtained both as a private network technology or as a public network service, or even as a hybrid. Early commercial implementations were towards deployment of Frame Relay service within the context of private corporate networks. However, there are a number of carriers that are currently providing public network Frame Relay services (both national as well as international services). One major application of Frame Relay is that it can be used for accessing WANs or for LAN interconnection. With the growth of LANs at the current rate, use of dedicated lines at T1 speeds for interconnecting LANs seems highly expensive and sometimes impractical especially when multiple remote sources are generating bursty traffic. Frame Relay service is a better alternative in such situations because it provides advantages such as sufficient capacity to support increasing bursty traffic in corporate environment, flexibility, survivability, universal access, resource sharing and control. Other applications of Frame Relay include block interactive data applications i.e. applications involving high resolution graphics, file transfer, multiplexed low bit rate and character interactive traffic. Frame Relay is not considered to be suitable for voice or steady-flow traffic.

The standardization of Frame Relay based transmission started in 1986. Two standards bodies, ITU-T and ANSI, have published various standards that specify the frame format, interfaces, and protocol stack for Frame Relay [70-73]. This section summarizes the current state of the performance management being addressed in these recommendations. The section starts off with a brief introduction of the frame format, interface and protocol stack associated with Frame Relay as specified in these standards and thereafter addresses the performance monitoring aspects of the Frame Relay networks.

2.6.1 Frame Relay Protocol and Protocol Architecture

2.6.1.1 Frame Relay Protocol Architecture

A Frame Relay network is typically composed of:

- user equipment supporting the frame relay interface
- one or more nodal processors (switches, concentrator, multiplexers etc.) owned by the user or a carrier, and
- communication links between the users and the nodal processors and between the processors.

The user equipment mainly consists of the appropriately configured LAN routers. A Frame Relay nodal processor is there to accept the frame it receives on one of its incoming ports, interpret the frame, segment it into cells (in some cases leave them as frames), and deliver it over the trunk connecting to a remote switch.

Frame Relay is a connection-oriented technology. A connection-oriented service involves a connection establishment phase, a data transfer phase, and a connection termination phase. Frame Relay can provide both PVCs (Permanent Virtual Circuits) and SVCs (Switched Virtual

Circuits). Currently, though, only PVCs are supported in most of the networks. In PVCs, frames traverse a fixed path through the network. This path is provisioned through a Network Management System long before the first call is made and remains in place for the requested duration. As a part of the provisioning process, the end points are specified and the network management system automatically builds a path between the nodes, and informs all the nodes in the network of the route. Although resources such as routing tables are updated immediately at the time of provisioning, bandwidth is allocated and dedicated only during the duration of the call, thus supporting bandwidth-on-demand.

In order to make the above described Frame Relay network a reality, a protocol stack has been specified that should be implemented in the user equipment as well as in the nodal processors to facilitate the Frame Relay based transmission [73]. This protocol is now known as the Link Access Procedure F-Core (LAP-F Core) protocol. In order to meet the requirements of the Frame Relay mode services, the data link layer is divided into the core sub-layer and the data link control sub-layer. The core sub-layer provides only those functions needed to take advantage of the statistical properties of the communications such as packet switching, frame mapping and multiplexing. The data link control sub-layer enhances the core sub-layer to support the OSI data link service. The nodes that terminate the end-user equipment over a link with a frame relay interface must also support CS (Convergence sublayer) and SAR (Segmentation and Reassembly sublayer) in order to segment the frames into cells, in case cell-based transmission is being used within the network. It should, however, be noted that the conversion of a frame into cells is not a part of Frame Relay specifications or requirements. This is needed in case the network uses cell-based transmission as its underlying transport mechanism.

2.6.1.2 Frame Relay Protocol

The Frame Relay Protocol specifies the format of the frames and the functionality supported by the protocol stack discussed above.

A frame, just like a packet, is a block of user data as created by the data link layer. It consists of a flag, a header, an information field, and a trailer. Different vendors of the data link layer software may produce frames of different format. The specific frame format to be used for Frame Relay services is defined in and is shown in Figure 2-4 [73].

The main fields in this format are the two flags, one at the beginning and the one at the end of the frame, the address field, the information field, and the frame check sequence field. The flags (1 octet each) are used as the delimiters. The flag, with 01111110 as its value, indicates the beginning as well as the end of a frame. The information field contains the information i.e. the user data or the administrative command/ response. This field could be anywhere from 1 octet to a default maximum size of 262 octets. ANSI specifies 4096 octets as the maximum, though 8198 octets may be used by some vendors. The frame check sequence is the checksum inserted by the sender and is used by the receiver to discover any errors in the frame data during transmission and, thus, to validate the frame.

The address field (2, 3 or 4 octets) contains C/R (command/response) indication, EA (Address Field Extension), DE (Discard Eligibility) indicator, BECN (Backward Explicit Congestion Notification), FECN (Forward Explicit Congestion Notification), and DLCI (Data Link Connection Identifier) bits. The size of DLCI bits may be 10, 17 or 23 depending upon the size of the address field being used. The EA bit indicates if the size of 2 octets, 3 octets or 4 octets is being used for the address field. The DE bit indicates whether the frame, in case of a network congestion, should be discarded or not. BECN and FECN bits are set by a congested network to indicate to the user that the congestion avoidance procedures should be initiated for the traffic in the opposite direction (or same direction) of the frame carrying the BECN (or FECN) indicator. DLCI identifies a virtual channel at a user to network or network to network interface. Consequently, a DLCI specifies a data link layer entity to/from which information is delivered/received and which is to be carried in frames by data link layer entities.

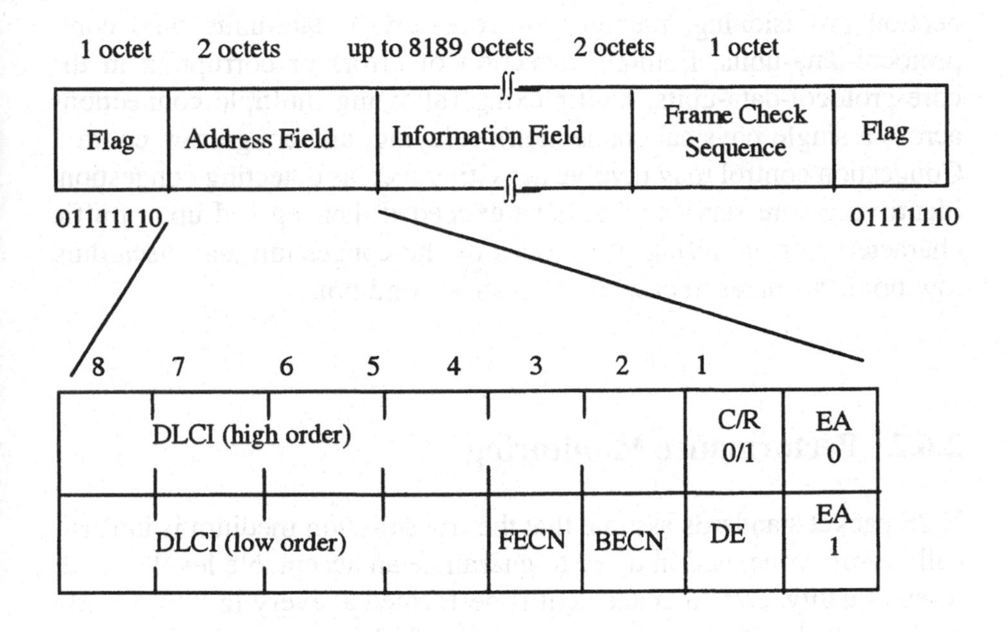

C/R: Command/Response indication.
EA: Address Field Extension bit
DE: Discard Eligibility indicator
BECN: Backward Explicit Congestion Notification
FECN: Forward Explicit Congestion Notification
DLCI: Data Link Connection Identifier

Figure. 2-4. Frame and Address Field Format

The Frame Relay protocol also outlines the functions [71] that should be provided by the Frame Relay protocol stack. The services provided by the core sub-layer to the data link control sub-layer include providing connections, exchange of core-service-data-units, providing core-connection-end-point-identifiers and controlling core service QoS parameters. The functions within the core sub-layer include core-connection provisioning, mapping of core-service-data-units onto core-protocol-data-units, framing, detection of errors or corruption in the core-protocol-data-units, multiplexing (allowing multiple connections across a single physical connection), relaying, and congestion control. Congestion control may involve activities such as detecting congestion, identifying core services that have exceeded their agreed upon traffic characteristics, notifying peer nodes of the congestion and discarding low priority frames in case of congestion condition.

2.6.2 Performance Monitoring

X.25 packet standards assume that the transmission medium is intrinsically error-prone, and in order to guarantee an acceptable level of end-to-end quality, error management is performed at every link by a fairly sophisticated but resource-intensive data link protocol. With a high-quality fibre-based communication infrastructure becoming a commonplace, many of the error correction and retransmission capabilities of X.25 can be safely eliminated. In Frame Relay, therefore, only error detection is done, while error correction and retransmission is completely eliminated and left for the end-user equipment. Since error correction and flow control are performed at the end-points, the Frame Relay expedites the process of routing packets through a series of nodes within the network. This causes reduction of bandwidth requirements and consequently the communication costs.

This section lists all the error conditions associated with the transmission performance of a Frame Relay service that are detected and reported by the Frame Relay protocol stack implemented at various nodes.

2.6.2.1 Performance Monitoring Locations

Performance monitoring is done at the Frame Relay nodes at the UNI as well as at the NNI.

2.6.2.2 Performance Monitoring Data

The error conditions detected by the core sub-layer include:

- Errored Frame: A Frame whose one or more bits are in error, lost, or extra (i.e. bits that were not present in the original signal). This condition is detected by Frame Check Sequence created using CRC-16.
- Lost Frame: A frame not delivered to the intended destination within the specified time-out period.
- Duplicated Frame: A frame received more than once by the destination.
- Mis-delivered Frame: A frame delivered to a destination other than the intended.
- Out-of-Sequence Frame: A frame received not in the same order as transmitted by the source.

The rates of the above mentioned error conditions would constitute the performance parameters for a Frame Relay service, for example errored frame rate, lost frame rate etc. As mentioned earlier, within the network, these frames in some cases are segmented into cells for cell-based transmission. If the underlying protocol used for this cell-based transmission within the network happens to be the ATM, then all the error conditions (and hence the performance parameters) associated with ATM based transmission, discussed in the last section, are also applicable to the associated Frame Relay based service.

2.7 Transmission Quality Assurance

This section discusses the process of verifying and controlling the quality of service. This process may be referred to as quality assurance, which consists of verifying that the intended network performance is achieved under day-to-day operating conditions, and of initiating appropriate action when performance is not as intended.

Broadly speaking, quality assurance includes procedures ranging from:

- manual, out-of-service (intrusive) testing of specific network components, to
- automatic and continuous in-service (non-intrusive) monitoring and centralized surveillance of networks and their components, to
- direct end-to-end subjective assessment based on user-opinion surveys.

While the subjective assessment is based on the input from the customers, the objective based maintenance can be of two types:

- Preventative Maintenance: which is defined as the detection of potential performance degradation before they affect service and are noticed by end users.
- Corrective Maintenance: which involves all the steps necessary from the moment that trouble is detected, to when the affected equipment has been repaired and returned to service.

The focus emphasis here is on the on-line monitoring of physical transmission systems for maintenance purposes, and on monitoring of digital paths for quality-of-service purposes.

2.7.1 Nature of Impairments

For a better understanding of functions related to quality assurance, it is useful to understand the nature of digital impairments. There are three principle classes of impairments as suggested in Figure 2-5, some or all of which may occur during the operation [90].

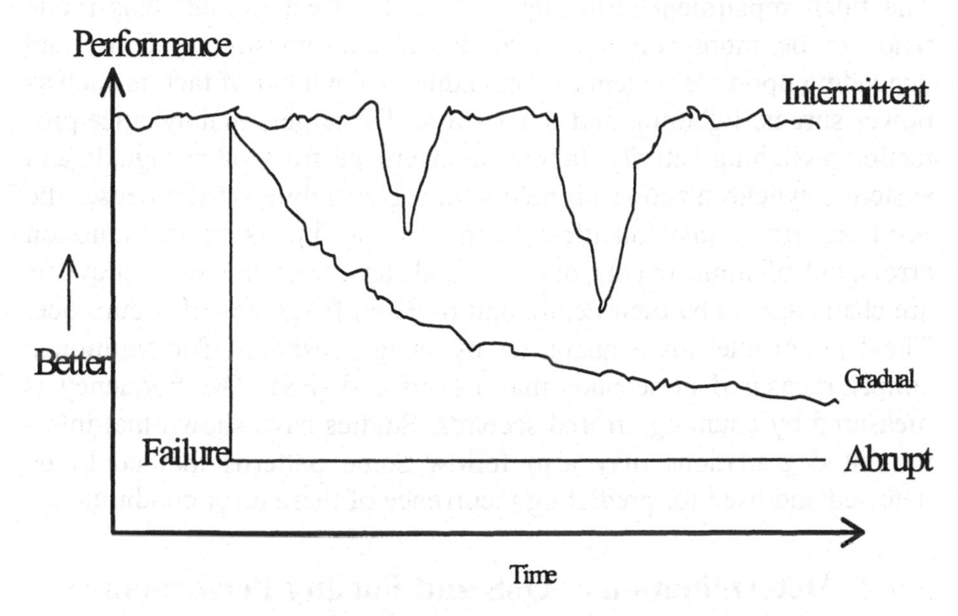

Figure. 2-5. Transmission Performance Impairments

The simplest mode is the hard/abrupt failures of which there are a number of different kinds, such as loss-of-signal (LOS) energy and loss-of-frame (LOF) synchronizations etc. Usually the maintenance response to these alarms is immediate corrective action either autonomously by NEs by switching to redundant capabilities, or by automatic/manual route restoration techniques to avoid or minimize loss of service.

The next impairment mode is the soft/gradual degradations. These degradations are consequence of cumulative events that may occur because of deterioration or drifting components in a NE, such as a timing recovery circuit in a regenerative repeater or line interface module. These deteriorations can be detected by counting errors or other events over time and producing alerts when thresholds are crossed, and consequently taking preventative or corrective actions as required depending upon the severity of the threshold that is crossed. An important aspect of the soft gradual degradation mode is that it lends itself to trend analysis techniques and to the application of preventative maintenance actions to avoid further degradations or failures.

The final impairment is the intermittent/transient events. This mode tends to be more common with digital transmission systems, and depending upon the system can arise due to a number of factors such as power surges, lightning and other static discharges, maintenance/protection switching activity, fading, interference from other signals and systems, synchronization problems, etc. Depending on the cause, the resulting errors may manifest themselves as bursts of transmission errors, out-of-frame events, or controlled slips. Such impairment events are characterized by their depth, duration and frequency of occurrence. The depth or intensity is quantified by using thresholds. The duration is simply measured in seconds that are errored (ES). The frequency is measured by counting errored seconds. Studies have shown that intermittent degradations may also follow some patterns that could be detected and used for predicting recurrence of these error conditions.

2.7.2 Determination of QoS and Facility Performance

The parameters identified in the pervious sections are standard measures of performance that are derived from the system specific performance primitives. The measure of each parameter in most cases is simply a count. For the purposes of data reduction, the fundamental measures are consolidated into one second bins. The second-by-second information is consolidated into 15 minute bins and passed on to OSs for analysis. CRC and BIP are the fundamental primitives, and ES and OOF are the fundamental measures of errors gathered at various points in the network. ESs are further categorized as ES-A, ES-B, and ES-C depending upon the intensity of the impairment.

In the context of layered performance monitoring, impairments propagate from functionally lower layers to the upper layer. For example regenerator section block errors cause transmission line block errors which in turn cause STS path block errors etc. The reverse is not true and therefore the lower layers can continue monitoring even if only the higher layer is in unavailable or error state.

Bit errors degrade service quality differently depending upon the type of the service. ES-A may relate to data transmission, ES-B to speech transmission and SES may represent a malfunction of the network elements. An event like OOF would cause a service disruption of at least

20 ms [87]. OOF events would very likely initiate protection switch causing an 8ms or 30ms (for automatic and manual protection switching, respectively) service interruption. These observed causal relationships between performance parameters and the QoS indicate that there is strong correlation between QoS parameters defined in [62][63] and the performance parameters identified in the previous sections. Various mathematical models have been derived that formulate the mapping between performance parameters such as ES and SES, and the QoS parameters [85][93][97]. These studies imply that the performance parameters, aggregated in 15 minutes bins, contain sufficient information to perform QoS conformance checking.

The performance parameters identified in the previous sections also reflect the performance of the underlying physical transmission systems. ES-A represents just the background noise, whereas, ES-C or SES and OOF represent a more serious problem (probably a malfunctioning in the facility), and may subsequently initiate protection switching activity.

2.7.3 Proactive Maintenance

The corrective maintenance relies mostly on the trouble detection through the surveillance of the service effecting alarms or may be based on the number and severity of performance monitoring alerts. It may also be based on the results of routine testing or from customer complaints. The preventative or proactive maintenance, however, involves the continuous surveillance and analysis of performance data to determine the extent of any degradations, and what subsequent preventative action should be taken.

Continuous monitoring thus provides a suitable infrastructure for preventative maintenance. The collected history of errors contains considerable knowledge to facilitate effective troubleshooting. Owing to the potential volume of data some form of automated data analysis is required to assess the extent and source of degradations, to predict impending more serious degradations through trend analysis, and perhaps to generate alarms in situations deemed to be serious enough to warrant corrective action. The preventative actions that can be taken in the event of significant performance degradations range from further diagnostic analysis to actual substitution of marginal, intermittent, or suspect equipment, so as to avoid a subsequent failure or further degradation. Other actions include manual or forced automatic protection switching to standby equipment or rerouting of performance sensitive traffic on to better transmission facilities [90].

There are a variety of maintenance philosophies relating to the analysis of performance data and alerts. The following section lists some of these data analysis techniques [85]. The overall intent of all these techniques or philosophies is to proactively address performance degradations before they affect end users of telecommunication services.

2.7.3.1 Thresholding

Thresholding is a simple way of determining when the performance of a network has exceeded a specified level. Low and high thresholds can be set to indicate the error conditions of two different severity levels, both of which are less severe than the alarm conditions. Consequently pro-active maintenance on the network can be performed by observing the threshold exceptions for degradation of performance.

2.7.3.2 Trending

The temporal variations in the gathered data can be used to determine the general trends in the performance of the network. It has been validated through observations that the ES-A demonstrate a trending and patterning capability. The 15 minute counts of ES-A, observed over several days, display trends leading to ES-B [87]. It has also been observed that there is a strong interdependence or correlation between SES and OOF events. SES, most likely, leads to an OOF situation over a period of few ms [85][93][97]. This information may be used to perform preemptive protection switching and avoid service disruption of longer periods from OOF situations.

The techniques that are most commonly used for this kind of data analysis are usually linear regression, Box-Jenkins ARIMA models, and may also include adaptive linear prediction techniques such as Kalman filters, LMS (Least Mean Square) filters and neural networks.

2.7.3.3 Pattern matching

Studies have shown that the transient degradations also show some patterns that could be exploited to predict or troubleshoot further degradations or faults [95][96][98]. The objective in these approaches is to observe the patterns of common behavior across circuits in the network using history data, and use the knowledge of network topology and network components, along with tests and measurements to determine the patterns that indicate that a problem will recur, estimate the severity of the problem, and correlate or sectionalize the problem.

2.8 Traffic Management

As mentioned earlier in the chapter, the performance objectives that deal with traffic are quite different from the performance objectives that deal with the transmission of this traffic. In the previous four sections, the performance monitoring of the transport resources was discussed. This section focuses on the performance monitoring of the switches, including traffic and quality of service measures.

The objective of traffic management is to enable as many calls as possible to be successfully completed. This objective is achieved by maximizing the utilization of resources. Traffic management is seen as a function of supervising the performance of the network and to be able, when necessary, to take action to control the flow of traffic, and to optimize the maximum utilization of the network capacity. The performance management OS will collect the performance information from the NE and send commands to modify its operations and or to reconfigure the network. In order for an NE to accomplish the above, it will need to perform the following [60]:

- collect traffic management information by the use of an internal measurement sub-system

- process traffic management information and convert it into traffic management indicators
- transfer the recognized set of traffic management indicators to the OS
- receive control information from the OS, and execute the appropriate controls to affect the traffic flow.

Traffic management is closely related to traffic and congestion control. Traffic control refers to the set of actions taken by the network in all relevant NEs to avoid congestion conditions or to minimize congestion effects and to avoid the congestion state spreading once congestion has occurred. However, congestion may occur, e.g. because of mis functioning of traffic control functions caused by unpredictable statistical fluctuations of traffic flows or of network failures. Therefore, additionally, functions referred to as congestion control functions are intended to react to network congestion in order to minimize its intensity, spread, and duration.

A wide range of traffic and congestion control functions have been specified to be used in the B-ISDN to maintain QoS of ATM connections [69]. These control functions have been listed below for information purposes and to help identify various parameters associated with these control procedures.

- Traffic Control functions
    - Network Resource Management (NRM)
    - Connection Admission Control (CAC)
    - Usage/Network Parameter Control (UPC/NPC)
    - Priority Control and Selective Cell Discarding
    - Traffic Shaping
    - Fast Resource Management
- Congestion Control functions
    - Selective Cell Discarding
    - Explicit forward congestion indication

Since Frame Relay is also a packet switching technology, its traffic and congestion control procedures are similar to ATM [72]. The ATM layer QoS is defined by a set of parameters such as delay and delay variation sensitivity, cell loss ratio, etc. The traffic parameters associated with ATM/B-ISDN (also Frame Relay) and hence the above listed control strategies include [69]

- peak cell rate
- average cell rate
- burstiness
- peak duration, and
- source type e.g. telephone, videophone.

The above set of control procedures and performance parameters are defined specifically for ATM/B-ISDN networks because of the packet-switched nature of ATM. However, other networks such as public-switched telephone networks (PSTN) and transport networks (SDH) also use similar control strategies for NRM, CAC and UPC/NPC. The remainder of this section lists the performance parameters and control strategies [60] associated with traffic management in PSTNs and SDH/SONET or T1/T3 based networks.

2.8.1 Network Status and Monitoring

This is a periodic or spontaneous collection of information about network status and traffic performance on digital resources of the telecommunication network.

The information could be provided as data, parameters and/or indicators. Network status information implies information about circuit groups as well as switching nodes.

2.8.1.1 Circuit Groups

- aspect of status information, e.g.
    - status of all circuit groups available to a destination
    - status of individual circuit sub-groups in circuit group

    - status of circuit in each sub-group
- aspects of the status indicators e.g.
    - when all circuits in a circuit group are busy
    - when all circuits in a circuit sub-group are busy
    - when all circuit groups to a destination are busy

2.8.1.2 Switching nodes

- load measurements: These are provided by call bids, usage or occupancy data, on the per cent of real time capacity available or in use, percentage of equipment in use, counts of second attempts, etc.
- Congestion measurement: These are provided by measurements of the delay in serving incoming calls, holding times of equipment, average call processing and set up time, queue lengths for common control equipment or software queues, counts of equipment time-outs, etc.
- Service availability of exchange equipment: This information will show when items of equipment are made busy for traffic.
- Congestion indicators: In addition to the above, indicators can be provided by digital exchanges which show the degree of congestion. These indicators can show:
    - no congestion (level 0)
    - moderate congestion (level 1)
    - serious congestion (level 2)
    - unable to process calls (level 3)
- Hard-to-Reach (HTR) Destinations: The HTR indicator reflects the status of the available/occupancy of the circuit groups of a particular destination (route)

2.8.1.3 Network Performance Monitoring/Control Requirements

Network performance information includes

- traffic on each circuit group

- traffic to each destination
- traffic to each digital exchange

Network performance data is generally expressed in parameters which help to identify difficulties in the network. Amongst these parameters are

- hard-to-reach (HTR)
- percentage overflow (% OFL)
- bids per circuit per hour (BCH)
- answer seizure ratio (ASR)
- answer bid ratio (ABR)
- seizures per circuit per hour (SCH)
- occupancy
- mean holding time per seizure
- busy-flash seizure ratio (BFSR)

The traffic managers can take the following control actions upon receiving the threshold crossing notifications for any of the above parameters.

- Protective Actions
    - temporary removal of circuits from service (circuit busying)
    - special instructions to operators
    - special recorded announcements
    - inhibiting overflow traffic, i.e. preventing traffic from overflowing onto circuit groups or into distant exchanges which are already experiencing congestion
    - inhibiting direct traffic: this action reduces the traffic accessing a circuit group in order to reduce the loading on the distant network
    - inhibiting traffic to a particular destination (code blocking or call gapping): This action may be taken when it is known that a distant part of the network is experiencing congestion

- Expensive actions
  - establishing temporary alternative routing arrangements in addition to those normally available
  - temporarily reorganizing the distribution of outgoing or incoming international traffic
  - establishing alternative routes into the national network for incoming international traffic
  - establishing alternative routes to an international exchange in the national network for originating international traffic.

2.9 Summary

An attempt has been made, in this chapter, to provide a framework for discussing performance management aspects of telecommunication networks. Key broadband telecommunication technologies have been discussed. The parameters that define the performance of these networks have been identified. The state-of-the art in performance management, specially performance monitoring in these networks has been presented by summarizing all the key functional requirements identified in various ITU-T documents.

Chapter 3

Performance Management in Cellular Networks

The subscription of cellular networks is continuing to grow at an unprecedented rate. Service providers need to accommodate this large subscriber base effectively so that a high return on investment could be realized. The solutions considered so far have mostly involved physical layer engineering and device intelligence. Network management, however, is also beginning to attract the attention it deserves. Network management procedures can significantly improve the resource utilization while maintaining an acceptable level of availability and quality of service. Moreover, the results of various management activities may also provide considerable insight into the design and planning of future 3G networks, thus enhancing the viability of the next generation of networks. For example, information on subscriber mobility and distribution, obtained for the existing wireless cellular networks, would be valid for future networks as well. Effectiveness of various network management procedures, however, lies on the performance monitoring abilities. Any optimization or diagnostic procedure must respond to an imminent or anticipated performance degradation in at least near real-time so that an acceptable level of service quality and availability could be maintained at all times. In this chapter, firstly a brief overview of various cellular technologies is provided. Thereafter, the performance management aspects of cellular networks are highlighted by presenting a performance management system for GSM/GPRS (Global System for Mobile Communications / General Purpose Radio System) networks. The system uses signaling data to estimate and manage the performance of the GSM/GPRS network. Some key performance parameters associated with GSM/GPRS networks are identified and the role of signaling data in deducing these parameters is pointed out.

3.1 Cellular Networks: an Overview

The radio spectrum available for cellular phone systems is limited. It is usually segmented into a number of voice channels. These channels are reused at regular distance intervals, providing spatial frequency reuse. The service area is typically divided into hexagonal shaped areas or cells. Each cell contains a BTS (Base Transceiver System), close to the geographic center, and is allocated a subset of channels. The same channel frequencies are reused in other cells, spaced sufficiently apart such that the co-channel interference is minimum. This frequency reuse or frequency planning is an important engineering task, requiring a good compromise between capacity and performance. Reducing the size of cells can improve the overall capacity, however the handover rate will increase.

After the call is accepted, the MS (Mobile Station) is assigned a channel through the BTS of the cell. The MS tunes to that channel and establishes communication. When the MS moves away from that BTS and tries to enter another cell, its signal strength deteriorates. Once the signal strength goes below a threshold, the handover procedure is initiated. An available channel from the target cell is then assigned to the MS. The MS tunes to this new channel and resumes communication. If all the channels in the new cell are already busy then the call is simply dropped. Network operators often minimize this call dropping by explicitly reserving some bandwidth for handover calls. Determining a suitable amount of bandwidth to be reserved for handovers is also part of the frequency planning process.

Each call involves at least two channels in each direction. One channel is the voice channel, and the other is the control channel. Voice channels carry user audio signals, whereas the control channel is used to exchange short data packets or messages such as location update messages and channels request messages from MS to BTS along with the channel assignment messages from BTS to MS. As soon as the MS is powered on, it scans the control channels and selects the strongest, usually from the nearest BTS. If it is a land-to-mobile call then the mobiles are paged over a paging channel (shared by all the MSs in the area) to indicate an incoming call. The target MS, if powered on, responds to the page and indicates its availability. It is subsequently assigned a free voice channel. On the other hand, for a mobile-to-mobile or mobile-to-land call, the MS sends a request over an access channel, again shared by all the MSs in the area through contention, containing its identification and the number it wants to reach. The network verifies its authentication, assigns a voice and a control channel to the MS, and also establishes the connection with the PSTN (Public Switched Telephone Network) in case the call is going to a land phone.

A cellular network is thus composed of three main resources: RF channels, terrestrial channels and signaling links. These are further elaborated below.

3.1.1 RF Channels

RF channels provide air interface between MS (Mobile Station) and PSTN. Various means of medium access are available including FDMA (Frequency Division Multiple Access), TDMA (Time Division Multiple Access), CDMA (Code Division Multiple Access), and random access schemes. FDMA implies frequency slotting i.e. each mobile station is assigned a piece of frequency band from the total available radio spectrum. A mobile station is tuned to this band at the call setup time and, thereafter, transmits in this frequency band. TDMA implies time slotting i.e. all MSs use the entire available radio spectrum, however, each mobile gets a time slot from the periodic time frame. The MS gets synchronized to this time slot at the call set up time and, thereafter, transmits only during this time slot at its peak rate. CDMA implies code slotting i.e. every user transmits at the same time using the entire available radio spectrum, however, the transmitted signal is modulated or multiplied by a unique code (orthogonal to every other user in the system). The base station demodulates or re-multiplies the received (aggregate) signal with that code to extract the user's signal. TDD (Time Division Duplex) and FDD (Frequency Division Duplex) schemes are also used. TDD is similar to TDMA and divides an FDMA channel into talk and listen slots to facilitate bidirectional communication. FDD is similar to FDMA and ensures that separate frequencies are used to transmit and receive.

In addition to the aforementioned controlled access schemes, some random access schemes are also used. Random access schemes are primarily used for packet data transmission. In these schemes any MS ready to transmit packet will do so, with some probability that collision will occur. In case of collision the packets are retransmitted, after waiting random amount of time. Examples include pure ALOHA and Slotted ALOHA protocols.

Some of the standards and their corresponding medium access techniques are listed below.

- The 1G standards:
    - AMPS (Advanced Mobile Phone System) is FDMA based; and
    - TIA's IS-88, IS-89 & IS-90 are N-AMPS (narrow-band AMPS) based standards.
- The 2G standards:
    - IS-54 is D-AMPS (Digital AMPS) standard based on TDMA/FDMA/FDD;
    - ETSI's GSM (Groupe Special Mobile, or, Global System for Mobile Communication) is based on TDMA/FDMA/FDD; and
    - cdmaOne employs CDMA for medium access.
- The 2.5G Standards include:
    - IS-95 is N-CDMA (narrow-band CDMA);
    - IS-136+ is an extension of D-TDMA; and
    - Phase II+ GPRS (General Packet Radio Service) is a random access based extension of GSM.
- The 3G or IMT-2000 Standards include:
    - cdma2000 is a replacement of IS-95;
    - UWC-136 is an extension of IS-136+; and
    - UMTS is a replacement of GSM/GPRS that will employ W-CDMA (Wideband-CDMA) technology.

Some of the packet radio standards are as follows:

- CDPD (Cellular Digital Packet Data)
  - It is a specification of the CDPD group. It uses infrastructure of existing cellular phone networks and hops on unused voice channels. Access to a channel is achieved through random access. Both, the connection-oriented as well as connectionless services are offered. Cell handovers, inter-system handovers and automatic nationwide roaming is supported.
- ARDIS (Advanced Radio Data Information Service)
  - It is used for bi-directional data transfer. An ARDIS specific infrastructure is used that includes base stations, radio network controller (RNC), and network control center (NCC). NCC controls packet routing, network management, billing and accounting. The customer's computer centers connect to the NCC via X.25 or SNA.
- Raw Mobile Data
  - It is a nationwide mobile data network and is based on Mobitex specifications. A Raw Mobile data specific infrastructure is used that includes NCC, main exchanges, area exchanges and base stations. Supports nationwide roaming.
- GPRS
  - GPRS is discussed later in this chapter.

3.1.2 Terrestrial Channels

Terrestrial channels are made available through a network of trunks providing connectivity among various NEs (Network Elements) of a cellular network, and POIs (Point Of Interface) between wireless providers and PSTN. T1/T3, DS1/DS3, OC-n, ATM or microwave links could be used. The performance management aspects of some of these technologies have been discussed in the previous chapter.

3.1.3 Signaling Channels

Signaling means the process of sending control information related to call setup, supervision and tear-down between network elements and databases. It also permits exchange of information required for distributed application processing and network management information. A signaling protocol defines the structure of information to be exchanged and the method of communication. Most of the current signaling systems are message/packet based.

There are basically three means of sending signaling messages:

- In-band Signaling. Examples include analog systems.
    - The signaling information is conveyed over the same channel that is used to carry speech.
    - The drawbacks include long call setup times (10-20 sec), limited information exchange and partial connected calls (resources are reserved on a link-by-link basis without verifying that the call will actually go through)
- Physical Out-of-band Signaling. Examples include SS7 (Signaling System 7).
    - A common but physically separate data channel is used to convey signaling information.
    - Advantages include high availability, better performance in call-control and adds to network intelligence.
- Physical In-band Logical Out-of-band. Examples include signaling between user and access node in ISDN where B-channel carries user-data and D-channel is mostly used to transport signaling (D-channel is also sometimes used to carry data traffic or short-messages).
    - Signaling and user traffic share the same physical channel but part of the capacity is exclusively reserved for carrying signaling traffic

Out-of-band signaling is also known as CCS (Common Channel Signaling) because a separate common channel is used for signaling.

In cellular networks, the signaling protocols used at the radio interface are part of the specific standard. For example, the signaling at the RF interface of GSM networks is GSM specific; though it is similar to that used in ISDN. On the other hand, the signaling network that connects a cellular switch i.e. MSC (Mobile Switching Center) to other MSCs, PSTN switches, or databases in the network is based on SS7. The examples include IS-41 (now ANSI-41) and GSM-MAP (Mobile Application Protocol). The IS-41 and GSM-MAP have gained tremendous importance for their ability to facilitate roaming through which a roaming subscriber can access a visited network just like its home network without any special operations/devices. The performance management aspects of SS7 are outside the scope of this discussion, however, the role of signaling in performance management of cellular networks is highlighted next using GSM/GPRS as an example.

3.2 Performance Monitoring in GSM/GPRS Networks

GSM/GPRS networks support both the circuit-switching as well as packet-switching. As a result, a broad spectrum of services are offered over these networks, supporting a large variety of QoS (Quality of Service) classes and traffic types. In order to provide services anytime and anywhere, multiple resources are consumed by each call, that include radio, terrestrial as well as signaling resources. The inherent multi-service & multi-resource nature of these networks, thus, implies that each call or connection will typically traverse multiple devices, switching technologies and transmission mediums. Even though the switches and routers used in these networks have significant intelligence, very-low-rate or intermittent errors or congestion conditions in these multiple equipment may interact, resulting in poor service quality. Continuous performance monitoring designed to either detect deterioration (leading to a reactive form of control) or detect characteristic patterns before service quality has dropped below an acceptable level (leading to a preventative form of control), is thus highly desired. Performance monitoring of multi-service & multi-resource GSM/GPRS networks, however, is anything but trivial. Large volumes of performance data from numerous locations in the network need to be collected to develop a comprehensive state of the network performance. Most of this information is usually collected by regularly reading the OM (Operational Measurement) registers or MIB (Management Information Base) variables maintained in network switches and routers. This approach has two main drawbacks. Firstly, the support for such activity imposes resource (CPU and memory) constraints on the NE (Network Element) being monitored. The resources that would otherwise be used for call or connection processing by the NEs are used for maintaining performance statistics for external performance monitoring systems. Secondly, during the times of extreme congestion or certain failure modes in the NEs, the support for such activity may become intermittent or unreliable. In other words, the performance data may not be available at times when it is needed the most.

Framework of a comprehensive performance management system is presented in this section that caters to the above implications and considerations. In the proposed system, named NOPM (Network Optimization Performance Manager), the aforementioned drawbacks of performance monitoring are alleviated by the use of signaling probes. These probes are attached to key signaling interfaces in the GSM/GPRS network, which are selected based on the hierarchical configuration of GSM/GPRS network. Flows of signaling messages being transported over these interfaces are monitored. Counts of various types of signaling messages or duration between specific signaling messages are interpreted to derive the performance parameters of interest.

3.2.1 Network Architectures and Signaling Protocol Stacks

The logical architecture of a GSM/GPRS network is as shown in Figure 3-1 [79]. A GSM network can be divided into three parts: The MS (Mobile Station), the BSS (Base Station Subsystem) and the NSS (Network & Switching Subsystem). The BSS is made up of the BSC (Base Station Controller) and the BTS (Base Transceiver Station). BTS provides the radio interface to the MS in a cell. It contains radio components including sender/receiver antennas for the cell. A group of BTSs are connected to a particular BSC which manages the radio resources for them. A cell is uniquely identified, in the service area, by its cell id and a LA (Location Area) code. A location area is a collection of contiguous cells used as a single entity for paging and user localization purposes. A BSC may have several LAs under its jurisdiction. The MSC (Mobile Switching Center) is the switching device, just like an exchange in the fixed network, but also provides all the functionality needed to handle a mobile subscriber. If the MSC also has a gateway function for communicating with other networks, it is called Gateway MSC (GMSC). In addition to these NEs the network also has some database types. The HLR (Home Location Register) database is used for maintaining subscriber information, and it also points to the current VLR (Visitor Location Register) of the subscriber. The VLR knows the current location of the MS and also keeps a copy of selected administrative information, obtained from HLR, about the subscriber currently in its domain. The EIR (Equipment Identity Register) contains a list of all the valid mobile station equipment within the network, where each mobile station is identified by its IMEI (International Mobile Equipment Identity). The authentication center (AuC) is a protected database that holds a copy of the secret key stored in each subscriber's SIM (Subscriber Identity Module) card located in the MS, which is used for authentication and encryption over the radio channel.

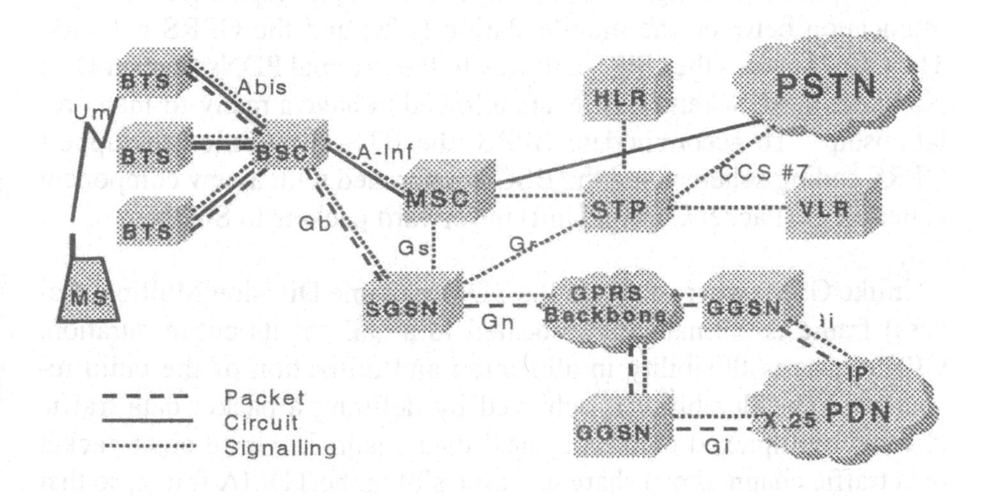

Figure. 3-1. Logical Architecture of a GSM/GPRS Network

The signaling between functional entities (registers) in the NSS employs MAP which is based on Signaling System 7 (SS7) [77]. SS7 is a packet-switched CCS (Common Channel Signaling) control network that is composed of SSP (Service Switching Point), STP (Signaling Transfer Point) and SCP (Signaling Control Point) components. SSPs are basically the MSCs and the PSTN switches. SCPs are the subsystems that maintain intelligence. HLR, VLR, EIR and AuC are some of the SCPs in the network. The signaling between an MS and a BTS, a BTS and a BSC, and a BSC and an MSC is point-to-point. The SS7 based signaling in the NSS, on the other hand, may involve a network of signaling links where each message may travel multiple hops, before reaching its destination. STPs, thus, provide the functionality of a router in the signaling network. Every SSP, STP and SCP in the signaling network is assigned a unique point code. A logical architecture of an SS7 network is shown in Figure 3-2.

GPRS maintains the same core GSM radio access technology, and provides packet data services by introducing two new packet-switching network elements called SGSN (Serving GPRS Support Node) and GGSN (Gateway GPRS Support Node) [81]. In addition, the GPRS register, which may be integrated with the GSM HLR, maintains the GPRS subscriber data and routing information. The SGSN is responsible for communication between the mobile station (MS) and the GPRS network. The GGSN acts as the GPRS gateway to the external PDNs (Packet Data Networks). SGSN and GGSN are allowed to have a many-to-many relationship. To accommodate GPRS, the BTS is modified to support GPRS coding scheme; and the BSC is upgraded with a new component called PCU (Packet Control Unit) to forward packets to SGSN.

Unlike GSM, where a slot in the TDMA (Time Division Multiple Access) frame is permanently allocated to a call for its entire duration, GPRS ensures flexibility in allocation and utilization of the radio resources. The flexibility is achieved by defining a packet data traffic channel multiplexed onto a physical data channel. Up to eight packet data traffic channels can share one time slot in the TDMA frame, so that eight MSs can share a single TDMA time slot for packet transmissions. Further, a multi-slot capability is also provided that allows transmission of data on multiple time slots (up to eight) within a TDMA frame, by a single connection.

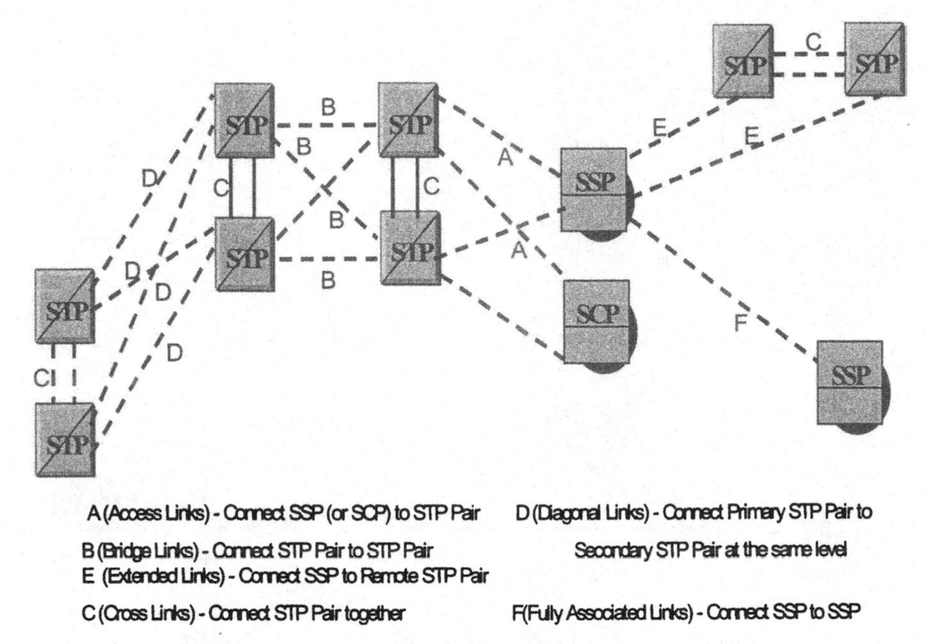

Figure. 3-2. Logical Architecture of an SS7 Signaling Network

3.2.2 The Proposed Framework

Figure 3-3 illustrates the functionality of NOPM, as embedded in the network operator's management infrastructure. The overall functionality involves performance data collection, trending/forecasting, performance modeling, report generation, GoS/QoS (Grade of Service / Quality of Service) evaluation and capacity optimization. The system can also interact with external fault and configuration management systems to achieve the performance objectives. The system can send notifications to a fault manager upon crossing of pre-specified performance thresholds or detection of abnormal events in the network. Similarly the results of various network optimization procedures could be passed on to a configuration manager or network provisioning system for implementation in the network [94][40].

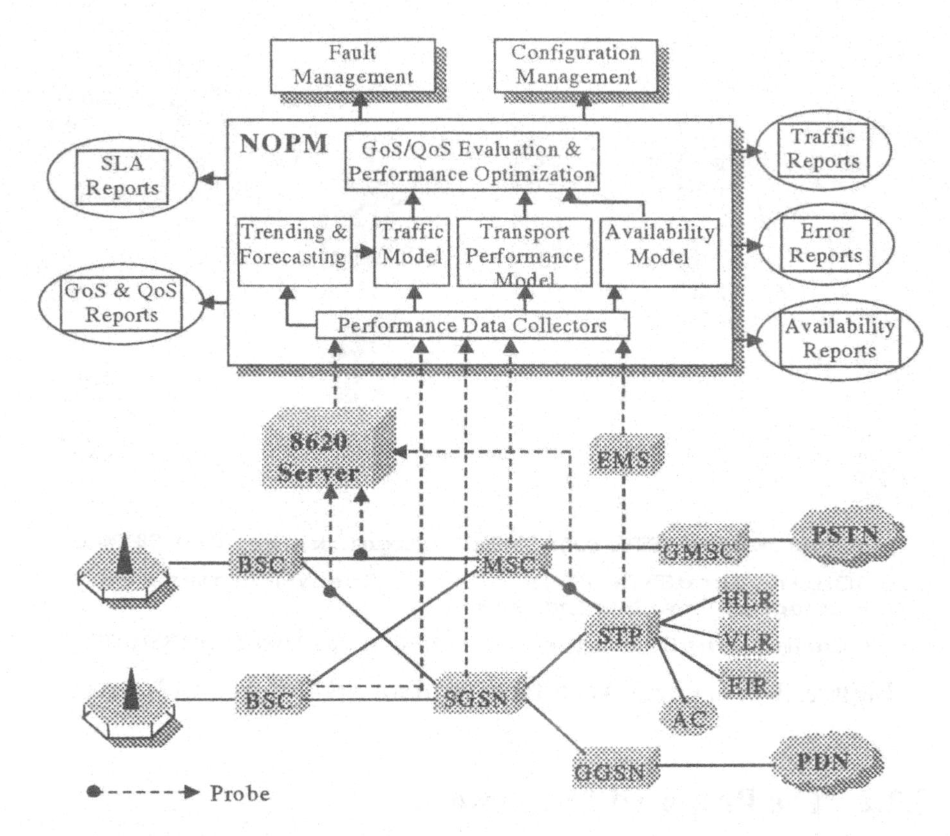

Figure. 3-3. Functional Framework of the Proposed System

As far as the performance monitoring is concerned, the main sources of data collection are the signaling probes. The signaling probes read messages belonging to all layers of signaling protocol stack on a signaling link in a GSM/GPRS network. These probes are controlled by a server. Various events, event counts and event duration, discovered through these signaling probes, are passed on to NOPM by the probe server. NOPM uses these measurements to deduce the network performance. The system also needs to collect some performance data directly from the NEs i.e. BSCs, MSCs, STPs, SCPs, SGSNs and GGSNs. These are mostly the measurements associated with the internal performance of the switches and routers, and are usually not exchanged over signaling links regularly. Examples include CPU utilization in a switch/router, size of available memory, and number of memory page faults etc. As mentioned earlier the data is collected by reading the switch's or router's OM registers or MIB variables. For this mode of data collection the devices are accessed through vendor specific CLI (Command Line Interface), EMS (Element Management System), or standards based SNMP or CMIP, if available. The data collection interval is user defined and is typically set to 5/15/30 minutes or 1 hourly, depending upon the network operations requirements.

3.2.3 GSM Performance Estimation

GSM offers circuit-switched services such as voice and HSCSD (High-Speed Circuit-switched Data). Packet based SMS (Short Message Service) is also supported but only through the signaling portion of the network. The BSS is the vital part of the GSM network. It controls the radio resources, which are at a prime in any wireless cellular network, due to the scarcity of the available radio spectrum. These resources must be utilized efficiently; and over-utilization as well as under-utilization of these resources must be detected in time to compute and implement an optimum frequency plan [94]. The BSS is also vulnerable to failures and errors. This is due to the fact that a relatively large number of devices and device types are involved in this section, making it more susceptible to failures caused by inter operability problems. The emphasis in this section is therefore on the monitoring of the BSS performance.

As apparent from the signaling plane of GSM depicted in Figure 3-4, the A-Interface is the prime location for monitoring BSS performance. This is due to the fact that the MSC is the main controlling entity in the GSM network [80]. Most of the radio resource management, connection management and mobility management operations are controlled by the MSC. Signaling Messages exchanged over this interface contain sufficient information to derive numerous parameters reflecting performance of network operations and resources at the BSS. However, additional information is sometimes needed that could be obtained by adding monitoring at the Abis-Interface. Signaling messages on Abis-Interface provide finer details of certain failures and abnormal events. Moreover, even though the MSC is the main switching device in the network, GSM standards allow BSCs to switch intra-BTS and intra-BSC handovers. In that case only the results of these operations are passed on to the MSC over the A-Interface. Thus, depending upon the type of BSCs used in the network, monitoring of Abis-Interface may be required for an accurate estimate of the offered load per cell.

Higher layers of the signaling protocol stacks, identified in Figure 2-4, are relevant to the subsequent discussions and need further elaboration here. The role of the RR (Radio Resource) management layer is to establish and release stable connection between MS and MSC for the duration of a call, and to maintain it despite user movements. The Mobility management (MM) handles the control functions required for mobility. The CM (Connection Management) is used to set up, maintain and take down calls connections. It is comprised of three subgroups e.g. CC (Call Control), SS (Supplementary Service) support and Short Message Service (SMS). Neither the BTS nor the BSC interpret CM and MM messages. They are simply exchanged between MSC and MS using the DTAP (Direct Transfer Application Part) protocol on the A-Interface. RR messages are mapped to or from the BSSAP (Base Station System Application Part) for exchange with the MSC. The BSSMAP protocol on the A-Interface is the counterpart to the RR protocol on the air interface.

Key performance parameters and the corresponding signaling messages are listed below. Each signaling message is identified by its signaling interface, protocol and direction on the interface.

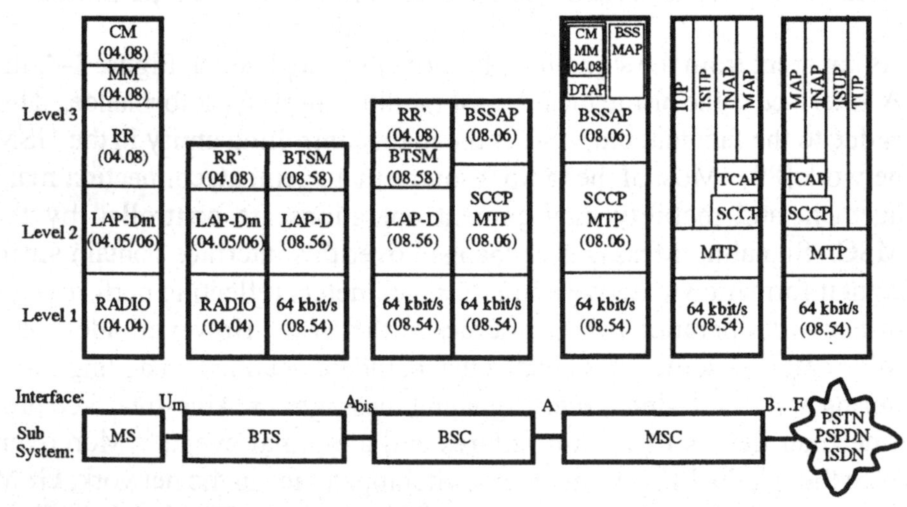

Figure. 3-4. GSM Signaling Plane

3.2.3.1 Resource Utilization

The parameters identified in this section are used to develop the call model, for each offered service, and the mobility model, for the associated subscribers. The call model implies statistics such as, average call rate per mobile, average call rate over entire service area, and average call duration time. In the wired telecommunication networks, the call model is sufficient to represent the traffic state of the network. In cellular networks, however, subscriber mobility significantly impacts the traffic characteristics, and needs to be taken into consideration. The obvious effect of subscriber mobility is the call handover. As soon as the mobile moves out of a cell, the channels occupied in the old cell are released, and the new cell then caters to the resource demand of this handover call. The channel occupancy time, in these networks, is likely to be less than the total call duration, as the mobile may be moving from cell to cell. The mobility model is, therefore, developed to represent the subscriber mobility characteristics through parameters such as, subscriber distribution, average call origination rate in a cell, average handover arrival rate in a cell and average channel occupancy time in a cell, etc. The call model and the mobility model are subsequently integrated for an accurate estimate of QoS as well as the capacity requirements of each service in each cell. The parameters are derived as follows:

Traffic Statistics	Corresponding Signaling Messages
Mobile Originating Call Attempts	Per target cell count of CM_SERVice_REQuest (A Interface, DTAP-MM, BSC to MSC) messages with CM_SERVice_REQuest_TYPE = 01_{hex}
Mobile Originating Calls Completed	Per target cell count of ALERTing (A Interface, DTAP-CC, MSC to BSC) messages
Mobile Terminating Call Attempts	Per target cell count of PAGing_ReSPonse (A Interface, DTAP-RR, BSC to MSC) messages
Mobile Terminating Calls Completed	Per target cell count of ALERTing (A Interface, DTAP-CC, BSC to MSC) messages

Incoming Handover Requests	Per target cell count of HaNDover_CoMmanD (Abis Interface, RR, BSC to BTS) messages with target cell different from serving cell (i.e. excluding intra-BTS Handover attempts)
Incoming Handover Completions	Intra-BSC Handover Completions + Inter-BSC Handover Completions + Inter-MSC Handover Completions
Average Cell Sojourn Time	This is applicable to MSs that either move into a cell or out of the cell during the progression of the call. For such MSs it is the average duration, computed per cell, between either the Incoming Handover Completion or Call Completion (Mobile Originating or Mobile Terminating) message whichever comes last, and either the subsequent Outgoing Handover Completion or the subsequent DISConnect (A Interface, DTAP-CC, MSC to/from BSC) message whichever comes last, of the calls
Average Call Holding Time	Average duration between the CONnect (A Interface, DTAP-CC, MSC to BSC) and the subsequent DISConnect (A Interface, DTAP-CC, MSC to/from BSC) messages with DISConnect CAUSE = 16 (i.e. normal clear), of the calls
Outgoing Handover Requests	Per serving cell count of HaNDover_CoMmanD (Abis Interface, RR, BSC to BTS) messages with target cell distinct from the serving cell (i.e. excluding intra-BTS handovers). Per serving cell count of HaNDover_CoMmanD (A Interface, BSSMAP, MSC to BSC) messages only provides a count of MSC controlled handovers that may not include BSC controlled handovers.
Outgoing Handover Completions	Per serving cell count of CLeaR_CoMmanD (A interface, BSSMAP, BSC to MSC) messages with cause $0b_{hex}$.
SMS Originations	Per target cell count of CM_SERVice_REQuest (A Interface, DTAP-MM, BSC to MSC) messages with CM_SERVice_REQuest_TYPE = 04_{hex}

Subscriber distribution, at various times of the day, reflects upon the variations in the underlying traffic load. This knowledge is useful for tracking variations in the capacity requirements, at various times of the day. Subscriber distribution is determined as follows:

Subscriber Distribution	Corresponding Signaling Messages
Location Update Requests	Per target cell count of LOCation_UPDating_REQuest (A Interface, DTAP-MM, BSC to MSC) messages

The traffic statistics derived above are used to compute the traffic load on each cell as follows:

Traffic Load	Estimation criteria
Average Channel Occupancy Time	1 / ((1 / Average Cell Sojourn Time) + (1 / Average Call Holding Time))
Offered Load	((Mobile Originating Call Attempts + Mobile Terminating Call Attempts + Incoming Handover Requests) / Observation Interval) · Average Channel Occupancy Time
Carried Load	((Mobile Originating Calls Completed + Mobile Terminating Calls Completed + Incoming Handover Completions) / Observation Interval) · Average Channel Occupancy Time

The traffic load measurements of each cell derived above also reflect the traffic load on the associated terrestrial links in the network. For example, the traffic load experienced by the trunk groups connecting a BTS to a BSC would be same as the traffic load on the associated cell. Similarly, the trunk groups connecting a BSC with an MSC would carry the load of all the BTSs homed to that BSC.

Various parameters that reflect the QoS of the network are determined as follows [94]:

QoS	Corresponding Signaling Messages
Calls Blocked due to congestion	Per target cell count of ASSingnment_FAILure (A Interface, BSSMAP, BSC to MSC) messages with ASSignment_FAILure_CAUSE = 21_{hex}
Handovers Blocked due to congestion	Per target cell count of HaNDover_FAILure (A Interface, BSS-MAP, BSC to MSC) messages with HaNDover_FAILure_CAUSE = 21_{hex}
Call Setup Delay	Average duration between the Call Attempt (Mobile Originating or Mobile Terminating) and the corresponding Call Completed messages of the calls
Call Termination Delay	Average duration between the DISConnect (A Interface, DTAP-CC, MSC to/from BSC) message with DISConnect CAUSE = 16 (i.e. normal clear) and the subsequent RELease_COMplete messages of the calls

3.2.3.2 Error performance

The transmission performance of a radio link in GSM is quantified in terms of RXQUAL (Received Signal Quality) and RXLEV (Received Signal Level) indicating the bit error rate and the interference level experienced by an MS on a channel, respectively. These measurements are read by the system from MEASurement_RESult (Abis Interface, RR, BTS to BSC) messages. A drop in the channel quality without any significant increase in the interference level is a sign of a degraded channel. In GSM this situation will automatically cause an intra-BTS handover. The system reports trends in temporal variations of the channel quality to assist in the cell engineering activities.

The transmission quality of trunks connecting BTSs, BSCs and MSCs is measured in terms of LOS (Loss of Signal), AIS (Alarm Indication Signal), LOF (Loss of Frame) alignments, BER (Bit Error Rate), CV (Coding Violation), CRC (Cyclic Redundancy Error Check) error count, Errored Seconds count (ES), Severely Errored Seconds count (SES), Unavailable Seconds count (UAS) etc. [62]. These measurements are obtained for a section, line, and path, depending upon the hierarchy of digital tributaries of the carrier. These parameters are, however, outside the scope of signaling probes and are monitored by querying the NEs (i.e. Add-Drop Multiplexers, Digital Cross-Connects etc.) directly.

As mentioned earlier, due to potential inter operability problems in the BSS, a call processing or mobility management activity is likely to experience failures at times, for reasons other than congestion. NEs then exchange signaling messages to inform the occurrence of such event. For example, an ASSignment_FAILure (Assignment Failure) message is sent from BSC to MSC over the A-Interface to indicate a resource assignment failure. The system reads these messages to identify the events and their causes. This information is reported by the system for diagnostic purposes. An assignment failure with cause "Requested Terrestrial Resource UnAvailable" or "Terrestrial Circuit Already Allocated" may suggest a mismatch in the trunk ids used in the BSC and the corresponding MSC, or perhaps a protocol error that caused an invalid trunk id to be used. The diagnosis could be further improved by including the corresponding the messages exchanged over the Abis-Interface. Again, as an example, in GSM, an assignment failure (at the radio interface) will generate an ASS_FAIL message on the A-Interface and also an ASS_FAI message on the Abis interface. While the cause in the ASS_FAIL message may be "Radio Interface Failure" or "Radio Interface Message Failure", the ASS_FAI message will provide further details such as "Radio Link Failure" or "Remote Transcoder Failure" to narrow down the root cause. Notable abnormal events with respective causes, and the associated signaling messages are listed below.

Error Condition	Signaling Message
Assignment Failure	ASSingment_FAIlure (Abis Interface, RR, BTS to BSC) and/or ASSingnment_FAILure (A Interface, BSSMAP, BSC to MSC) message.
Incoming Handover Failure	HaNDover_FAILure (Abis Interface, RR, BTS to BSC) message
Clear TCH (Traffic Channel) Request	CLeaR_CoMmanD (A Interface, BSSMAP, MSC to BSC) messages that request clearing of a TCH due to premature termination of a call.

Clear SD-CCH (Stand Alone Control Channel) Request	CLeaR_CoMmanD (A Interface, BSSMAP, MSC to BSC) messages that request clearing of an SDCCH due to premature termination of a call.
Handover Required	HND_RQD (A Interface, BSSMAP, BSC to MSC) message.
Failure Causes in messages on the A-Interface.	Radio Interface Message Failure, Radio Interface Failure, O&M Intervention, Radio Interface Failure – Reversion to Old Channel, Equipment Failure, No Radio Resource Available, Requested Terrestrial Resource UnAvailable, Requested Transcoding/Rate Adaptation Unavailable, Terrestrial Circuit Already Allocated, Uplink Quality, Uplink Strength, Downlink Quality, Downlink Strength, Distance, Response to MSC Invocation, and Better Cell etc [78]
Rejection of Location Update	LOCation_UPDating_REJect (A Interface, DTAP-MM, MSC to BSC) message. The possible causes include HLR Not Reachable and Unknown IMSI/TMSI etc.
Connection Failure	CONNection FAILure (Abis Interface, RR, BTS to BSC) message is an indication of Layer 1 errors on the Air Interface. The possible causes include Radio Link Failure, Handover Access Failure and Remote Transcoder Failure etc.
Error Indication	ERRor_INDication (Abis Interface, RR, BTS to BSC) message is an indication of layer 2 errors on the Air Interface. The possible causes include Frame Not Implemented and Timer T200 Expired etc.

Abnormal terminations of calls due to failure events is an important QoS parameter and is derived as under:

QoS	Signaling Message
Calls Cleared	Count of Assignment Failures and Handover Failures with reason other than "No Radio Resource Available"

The duration of the failure modes, identified above, is used to determine availability, which is the percentage of time a network component or function is operational. It is computed using measurements such as MTBF (Mean Time Between Failures) and MTTR (Mean Time To Recover). GSM also conducts maintenance operations during which certain functions and components are unavailable. Duration of these downtimes is also included in the availability computations. Certain OAM (Operations Administrations and Maintenance) messages are exchanged in GSM to indicate the commencement and termination of these operations. The system uses these messages to determine the mean downtime of the associated maintenance activities. For example the duration between BLOck/CIRCuit_GrouP_BLOck (A Interface, BSSMAP, MSC to/from BSC) and subsequent UnBLOck/ CIRCuit_GrouP_UnBLOck (A Interface, BSSMAP, MSC to/from BSC) messages would indicate the downtime of the specified trunk or trunk group.

3.2.4 GPRS Performance Estimation

GPRS supports data services such as SMS, E-mail, file transfer and web access. The signaling plane of GPRS is depicted in Figure 3-5 [81]. The Gb-Interface in the GPRS signaling plane that connects BSS to SGSN is the counterpart of A-Interface. The protocols that need to be defined for clarity in the subsequent discussion are the higher layers of the protocol stack at the Gb-Interface. The GMM/SM (GPRS Mobility Management / Session Management) protocol supports mobility and session management functionality such as GPRS attach, GPRS detach, security, routing area update, location update, PDP context activation, and PDP context deactivation. The LLC (Logical Link Control) provides highly reliable and ciphered logical links [82]. LLC (Logical Link Control) PDUs (Packet Data Units) transport signaling messages as well as data packets [83]. BSSGP (BSS GPRS Protocol) conveys routing and QoS related information between BSS and SGSN. BSSGP does not perform error correction but provides virtual connections to LLC PDU flows. Each connection is assigned a BVCI (BSSGP Virtual Circuit Identity). Unlike A-Interface, where a trunk or a physical channel is exclusively assigned to a call for its entire duration, all active connections as well as signaling flows at the Gb-Interface share the same resources, and therefore, the same probe monitors signaling and data transmission between BSS and SGSN. Key GPRS performance parameters are listed in the following sections along with the signaling messages used to derive them.

3.2.4.1 Resource Utilization

An MS must first attach itself to GPRS and activate a PDP (Packet Data Protocol) context to be able to initiate the data transfer. The PDP context contains QoS requirements and transmission rates of a session, besides other parameters. Various QoS classes that are available through GPRS are: Precedence Class with values {high, medium, low}; Delay Class with values {0.5s, 5s, 50s, best effort}; and Reliability Class with values {1-5} [81]. A session can transmit with possible peak throughput of {1, 2, 4, 8, 16, 32, 64, 128, 256} KOctet/s and mean throughput of {.1, .2, .5, 1, 2, 5, 10, 20, 50, 100, 200, 500, 1000, 2000, 5000, 10000, 20000, 50000 KOctets/hr, or best-effort}. The QoS and transmission rate information contained in the PDP contexts is used for CAC (Call Admission Control) at the SGSN. A PDP context at any given time may be in an active state or an inactive state at the SGSN. Thus, a GPRS attach request is equivalent to a connection attempt, whereas the request for PDP activation is the request for transmission within the session.

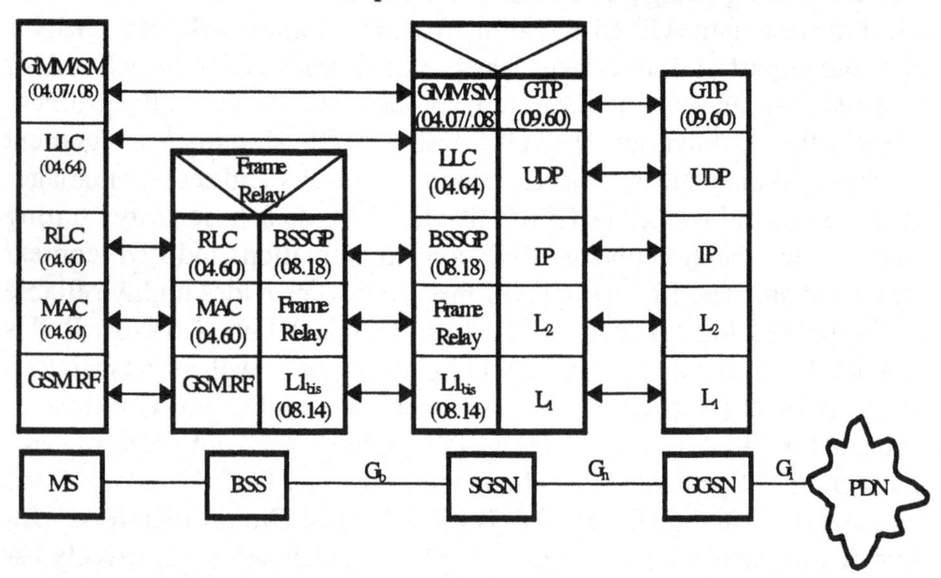

Figure. 3-5. GPRS Signaling Plane

The mobility contexts are maintained at the MS as well as SGSN. A mobility context may exist in one of the three possible states i.e. idle, standby and active. The MS can transmit and receive data only in the active state. While in this state, the MS must perform a location update upon entering a new cell, referred to as a cell update. In case no packet is received by SGSN for a certain period of time, a time out occurs causing a transition of mobility context state from active to standby. In standby state, the MS is required to perform a location update upon entering a new RA (Routing Area), referred to as RA update. A routing area is a collection of cells, used as a single entity for routing and paging purposes by SGSN. An SGSN may have multiple RAs under its jurisdiction. The MS cannot send or receive data in the standby state; though it can receive page requests in this mode. A valid page response from MS, then, causes the transition to the active state. In an idle state the MS can perform cell selection and LA based location updates, but is otherwise unknown to GPRS. The key traffic and utilization parameters along with the corresponding signaling messages are listed below.

Traffic	Signaling Messages
Mobile Connection Attempts	Per target cell count of Attach_Attempt (G_b Interface, GMM, MS to SGSN) messages
Mobile Connections Completed	Per target cell count of Attach_Accept (G_b Interface, GMM, SGSN to MS) messages
Mobile Originating Transfer Attempts	Activate_PDP_Context_Request (G_b Interface, GPRS SM, MS to SGSN) messages.
Mobile Originating Transfer Attempts Successful	Activate_PDP_Context_Accept (G_b Interface, GPRS SM, SGSN to MS) messages.
Mobile Terminating Transfer Attempts	Per target cell count of GPRS Paging_Request (G_b Interface, GMM, SGSN to MS) messages.
Mobile Terminating Transfer Attempts Successful	A Paging_Request is considered to be accepted by the MS if a valid LLC PDU (carried inside a BSSGP PDU) is received from the MS.

Subscriber Distribution:	Cell based: A cell update is any correctly received and valid LLC PDU carried inside a BSSGP PDU containing a new identifier of the cell. RA (Routing Area) based: Per target RA count of Routing_Area_Update_Request (G_b Interface, GMM, MS to SGSN) messages. LA (Location Area) based: Per target RA count of Routing_Area_Update_Request (G_b Interface, GMM, MS to SGSN) messages with Update_Type = 1 or 2
Requested QoS Profile of the Connection	Activate_PDP_Context_Request (G_b Interface, GPRS SM, MS to SGSN) message. The IE (Information Element) of interest is the Requested_QoS in the Activate_PDP_Context_Request PDU.
Accepted QoS Profile of the Connection	Activate_PDP_Context_Accept (G_b Interface, GPRS SM, SGSN to MS) message. The IE of interest is Negotiated_QoS in the Activate_PDP_Context_Accept PDU.
Average duration a PDP Context is active	Average duration between the Activate_PDP_Context_Accept (G_b Interface, GPRS SM, SGSN to MS) and, either the subsequent Detach_Request (G_b Interface, GMM, SGSN to MS) or DeActivate_PDP_Context_Request (G_b Interface, GPRS SM, SGSN to/from MS) message, whichever comes first.
Average Connection Time	Average duration between the Attach_Accept (G_b Interface, GMM, SGSN to MS) and the subsequent Detach_Request (G_b Interface, GMM, SGSN to/from MS) messages.
Average Residence Time in the RA	Average duration between two consecutive Routing Area Update messages with target RA specified in the first one.
Average Residence Time in the Cell	Average duration between two consecutive Cell Update messages with target Cell specified in the first one.
Downlink LLC PDUs Transmitted	Per connection or target cell count of DL-UNITDATA (G_b Interface, BSSGP, SGSN to BSS) PDUs
Uplink LLC PDUs Transmitted	Per connection or target cell count of UL-UNITDATA (G_b Interface, BSSGP, BSS to SGSN) PDUs

Various parameters that reflect the QoS in a GPRS network are derived as follows.

QoS	Corresponding Signaling Messages
Attach Requests Blocked due to congestion	**Per target cell count of Attach_Reject (G_b Interface, GMM, SGSN to MS) messages with cause = 32 (i.e. congestion)**
Connection Set-Up Delay	**Average duration between an Attach_Request (G_b Interface, GMM, MS to SGSN) and a subsequent Attach_Complete (G_b Interface, GMM, MS to SGSN) message.**
Connection Termination Delay	**Average duration between a Detach_Request (G_b Interface, GMM, MS to/from SGSN) and a subsequent Detach_Complete (G_b Interface, GMM, MS to/from SGSN) message.**
Onset of Congestion	**STATUS (G_b Interface, BSSGP, BSS to/from SGSN) message with Cause = 06_{hex} (Cell Traffic Congestion) and 07_{hex} (SGSN Congestion)**
Transmission Blocked due to congestion	**Per BVCI or target cell count of Activate_PDP_Context_Reject messages with Cause=26 (i.e. congestion) and Deactivate_PDP_Context_Request messages with cause = 37 (i.e. QoS not Accepted)**
Average Queuing Delay	**FLOW_CONTROL_BVC (G_b Interface, BSSGP, BSS to SGSN) message. The delay value (centi-seconds) in the BVC_Measurement IE indicates the average queuing delay that needs to be assured by the BSSGP flow control mechanism for the specified BVC.**

3.2.4.2 Radio Resource Error Performance

The transmission performance of a GPRS session at the radio interface is determined through RADIO_STATUS (G_b Interface, BSSGP, SGSN(BSS) message [80]. The possible conditions in the status field of the RADIO_STATUS PDU are Radio Contact Lost with the MS; Radio Link Quality Insufficient to Continue Communication; and Cell-Reselection Ordered.

Various failure and abnormal conditions associated with GMM/SM are listed below, along with the corresponding signaling messages.

Error Condition	Signaling Message
Connections Denied	Attach_Rejects (G_b Interface, GMM, SGSN to MS) messages
Failure Causes of various GMM Operations [78]	IMSI unknown in HLR, Illegal MS, Illegal ME, GPRS services not allowed, GPRS services and non-GPRS services not allowed, MS identity cannot be derived by the network, Implicitly detached, PLMN (Personal Land Mobile Network) not allowed, Location Area not allowed, Roaming not allowed in this location area, GPRS services not allowed in this PLMN, MSC temporarily not reachable, Network failure, Congestion, To retry upon entry into a new cell, Semantically incorrect message, Invalid mandatory information, Message type non-existent or not implemented, Message type not compatible with the protocol state, Information element non-existent or not implemented, Conditional IE error, Message not compatible with the protocol state, Protocol error – unspecified
Transmissions Rejected	Activate_PDP_Context_Rejects (G_b Interface, GPRS SM, SGSN to MS) messages
Failure Causes of various GPRS SM Operations [78]	Insufficient resources, Missing or unknown APN (Access Protocol Name), Unknown PDP address or PDP type, User Authentication failed, Activation rejected by GGSN, Activation rejected, unspecified, Service option not supported, Requested service option not subscribed, Service option temporarily out of order, NSAPI (Service Access Point Id) already used, Regular deactivation, QoS not accepted, Network failure, Reactivation required, Feature not supported, Invalid transaction identifier value, Semantically incorrect message, Invalid mandatory information, Message type non-existent or not implemented, Message type not compatible with the protocol state, Information element non-existent or not implemented, Conditional IE error, Message not compatible with the protocol state, Protocol error, unspecified Service option temporarily out of order, and Protocol error – unspecified etc.

BSSGP Abnormal Conditions [80]	Abnormalities in BSSGP are detected through STATUS (G_b Interface, BSSGP, BSS(SGSN) PDU. The possible causes include Processor overload, Equipment failure, Transit network service failure, Network service transmission capacity modified from zero kbps to greater than zero kbps, Unknown MS, BVCI (BSSGP Virtual Circuit Identifier) unknown, cell traffic congestion, SGSN congestion, O & M intervention, BVCI-blocked, PFC create failure, Semantically incorrect PDU, Invalid mandatory information, Missing mandatory IE, Missing conditional IE, Unexpected conditional IE, Conditional IE error, PDU not compatible with the protocol state, Protocol error – unspecified, PDU not compatible with the feature set, etc.

Key QoS parameters are determined as follows:

QoS	Signaling Message
Abnormal Connection Terminations	Count of Attach_Reject and Activate_PDP_Context_Reject messages per connection or cell
LLC PDUs Discarded [80]	LLC-DISCARDED (G_b Interface, BSSGP, BSS to SGSN) message indicates that a number of buffered LLC-PDUs in a cell for an MS have been deleted inside the BSS (because of PDU Lifetime expiration or Cell-reselection for example)

Again, as in GSM, duration of failure modes as well as maintenance activities are used to compute the availability of GPRS components and operations. The duration of maintenance events is computed using OAM signaling messages. Examples of OAM signaling messages in GPRS include BVC_Blocked, BVC_Unblocked and BVC_Reset that indicate blocked, unblocked and reset state of the specified BVC, respectively [80].

3.3 Summary

Application of GSM/GPRS signaling in network performance monitoring is highlighted. A thorough study has been conducted to demonstrate that a small set of signaling data could be effectively interpreted to estimate the performance of various network operations and resources in a GSM/GPRS network. Not only the signaling performance is measured but also the radio as well as the terrestrial resources are measured. Framework of a performance management system, NOPM, has been presented that exploits the aforementioned aspect of GSM/GPRS signaling in determining the overall network performance. Signaling probes are used to monitor signaling message flows at key signaling interfaces of the GSM/GPRS network. A select group of signaling events, event counts and event duration is thereafter used to measure the network performance. The network performance is expressed in terms of traffic load measurements, resource utilization measurements, transmission quality, QoS, availability measurements and failure event distributions. The performance data is collected, processed and reported in at least near real-time to facilitate effective network optimization and diagnosis.

Chapter 4

Estimation & Prediction of Multiplexed VBR Traffic

The capability to predict VBR traffic can significantly improve the effectiveness of packet level control in high-speed packet-switched networks. In this chapter, the feasibility of estimation and prediction of multiplexed VBR video traffic is demonstrated.

Firstly, multiple video-conferencing and video-phone sources employing the DCT for intra-frame compression with no motion compensation are considered [37, 38]. Time-invariant least mean-square filters are proposed for estimating and predicting multiplexed traffic from such sources. The source traffic is modeled as an AR(1) modulated process, with model parameters assumed to be known. Using this source model, the first and second order statistics of the multiplexed traffic are computed. These statistics are used for determining the coefficients of filters used for estimation and prediction. The performance of the filters is evaluated by using synthetic data.

Secondly, broadcast quality video sources employing MPEG compression are considered [39]. Time sequenced adaptive filters are proposed for linear prediction of multiplexed traffic from such multiple sources. The source traffic is modeled as a concatenation of PAR (Periodic AutoRegressive) cyclostationary processes with model parameters unknown. Time sequenced adaptive filters allow for the cyclostationary nature of the input by periodically changing the filter and adaptation parameters. This predictor set up is ideal for predicting multiplexed MPEG traffic that has periodically recurring statistical properties and can be considered as a concatenation of PAR (Periodic

AutoRegressive) cyclostationary processes. The viability of the approach is illustrated using a number of half-hour long empirical MPEG-1 traces. Both NLMS (Normalized Least Mean Square) as well as RLS (Recursive Least Squares) algorithms are considered for adaptation. The results indicate that the RLS algorithm clearly outperforms the conventional NLMS adaptive algorithm in terms of convergence speed and steady-state mean-square prediction error; and, hence, is a more suitable candidate for such an application.

Lastly, assuming that the call arrival process is Poisson with exponential call holding time, the probability of a new call arrival is determined using the transient-state analysis of a birth-death Markov Process. These predicted values could be combined to predict the onset of congestion.

As mentioned in Chapter 1, the feasibility of predicting VBR traffic using linear predictors has also been explored previously in [1], [8], [21], [52], and [56]. Adas [1], for example, uses an LMS adaptive filter to separately predict I, P, and B frame series of an MPEG-1 trace. The approach proposed in this chapter, however, is distinct from the aforementioned work in three aspects. Firstly, unlike Adas [1], the proposed approach has been developed to predict the multiplexed traffic instead of a single source. As a result, the approach by Adas [1] requires the knowledge of frame boundaries in order to isolate, and then predict, different types of frames from a single source. The proposed approach, in contrast, estimates and predicts the multiplexed traffic without requiring any knowledge of the frame boundaries by using small observation windows and taking the intra-frame packet arrival process into consideration. Secondly, in Adas [1], the I, P and B frame series are predicted separately and, therefore, the crosscorrelation among these frames, as observed by Doulamis *et al.* [13] and Lambardo *et al.* [27], is not included for prediction. On the other hand, in the proposed approach, the predictor takes into account the autocorrelation as well as the crosscorrelation among the frames. Lastly, Adas [1], does not consider predicting the time of arrival of a new call. In the proposed approach the time of arrival of a new call is also predicted, which makes this approach more robust.

4.1 The System Model

A typical ATM node can be represented as in Figure 4-1. It is a statistical multiplexer receiving packet streams of digitized VBR video from a maximum of S number of sources. The sources are assumed to be unsynchronized and uncorrelated. Let $s = s(t) \in \{1 \ldots S\}$ be the number of sources in an active state at time t. Considering that the network queueing and the packetizing effects can be ignored for the VBR traffic [51], the output of the statistical multiplexer is an aggregate of the inputs from the s active sources. The task, therefore, is to measure the intensity of cell (the fixed size ATM packets are commonly referred to as cells) transmission at the multiplexer output during a small observation window and use it to estimate and predict the aggregate future transmission rate at the output channel. In other words using the cell count $c(j)$ in the jth observation window and the cell counts in some preceding observation windows an estimate $\hat{x}(j)$ of the actual output transmission rate $x(j)$ is obtained, and thereafter $\hat{x}(j+k)$ is obtained given $\hat{x}(j)$. Also, given that $s(t) = s$, some future state $s(t+\tau)$ of the activity of VBR sources is predicted. As we shall see, these predicted values can then be combined to forecast the onset of congestion.

Filters used for estimation and prediction of multiplexed VBR traffic, depending upon the underlying source traffic, are described next.

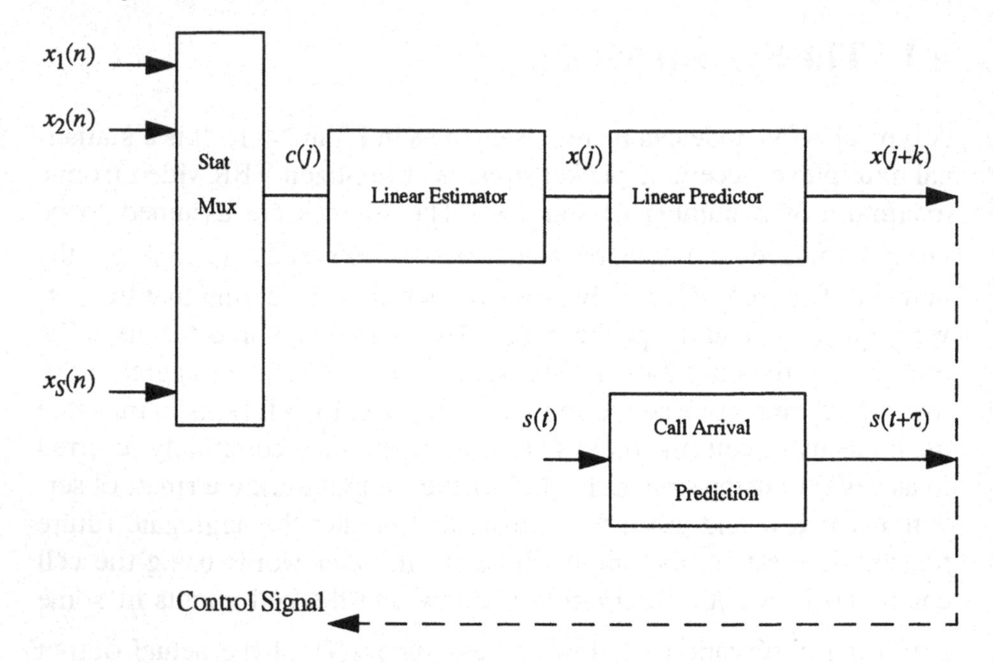

Figure. 4-1. Model of the Proposed Scheme.

4.2 Video-Conferencing or Video-Phone Sources

Each source in Figure 4-1 is assumed to be a video-conferencing or a video-phone source employing the DCT for intra-frame compression with no motion compensation. Thus, when active, a source i generates VBR traffic as

$$x_i(n) = \alpha x_i(n-1) + \beta w_i(n) \tag{4.1}$$

where $x_i(n)$ is the size of nth frame in bits/pixels or bits/frame, $w_i(n)$ is i.i.d. additive white Gaussian noise with mean 0.572 and unity variance, and α and β are the two known constants [31]. The index n is the frame number. The empirically estimated mean, variance, and the autocorrelation function of a VBR traffic trace are equated with the mean, variance and the autocorrelation function, respectively of the AR(1) process of (4.1) to obtain the model parameters, as described in [31]. The process $x_i(n)$ has Gaussian probability density and an exponential autocovariance function given as [31]

$$r_{x_i x_i}(m) = E\{x_i(n)x_i(n+m)\}$$

$$= 0.0536 \times e^{-0.13m} \text{ (bits/pixel)}^2. \qquad (4.2)$$

The AR(1) process of (4.1) represents the inter-frame variations in the bit rate of the VBR source or in other words represents an AR(1) process sampled at every T sec where T is 1/30 sec assuming 30 frames/sec. As the frame boundaries of the VBR sources in Figure 4-1 may not always be synchronized, the observation window size should be very small so that there is a negligible likelihood of a frame transition during this small observation interval by any of the sources. For the observation windows smaller than T, the intra-frame cell arrival process needs to be considered as well.

Once a video frame is scanned and subsequently compressed by the video encoder, the information bits (or packets of bits) are transmitted to the multiplexer. The information bits or packets can arrive in a burst, as Poisson distributed, as uniformly distributed, or in a deterministic fashion [12, 47]. Burst mode is unsmoothed while the Poisson, Uniform and Deterministic are smoothed modes of arrival. In the Burst mode, the sources are either ON at a known peak rate or OFF. In the Poisson mode, the packets arrive with exponentially distributed inter-arrival times. In the Uniform mode, packets arrive at random uniformly distributed over the entire frame. It has been demonstrated by Dixit and Skelly [12] that within the frame interval the packet arrival times are approximately exponentially distributed and, thus, this mode could be approximated to Poisson arrival mode. In the Deterministic mode, the source generates cells with the deterministic spacing and the data is exhausted by the end of the frame period.

It has been contended in literature that the Burst arrival mode is highly in-efficient from the multiplexer performance point-of-view and these intra-frame high energy fluctuations of the bit rate should be smoothed out either at the codec or through pre-buffering [47, 48, 51]. Thus the intra-frame arrivals are assumed to be smooth. We rewrite (4.1) in the continuous form as

$$\tilde{x}_i(t) = \sum_{n=0}^{\infty} x_i(n)h(t-nT) \tag{4.3}$$

where $x_i(n)$ is the AR(1) process of (4.1) and $h(t)$ is a rectangular function of width T. It is quite important to note that it is the inter-frame bit rate that varies according to an AR process which can be predicted. The instantaneous bit rate, computed by using cell counts over much smaller observation windows, may be noisy with respect to $x(n)$, because of the effects of burstiness over smaller observation windows or measurement errors. The objective of the estimator, therefore, is to minimize the impact of this noise and obtain a best possible estimate of $x(n)$. This section proposes a traffic model for the multiplexer output and lays proper foundation for determining the response of such an estimator.

The process $\tilde{x}_i(t)$, in (4.3), is a cyclostationary process, at least in the wide-sense, as its mean and autocorrelation function are periodic with a period T [35]. However, assuming that the arrivals are unsynchronized and thus have a random phase, uniformly distributed between $[0,T]$, makes the shifted process $x_i(t) = \tilde{x}_i(t-\theta)$ into a stationary process. The mean and the autocovariance of this process are then given as

$$E\{x_i(t)\} = \frac{1}{T}\int_0^T E\{\tilde{x}_i(t)\}dt = E\{x_i(n)\} \tag{4.4}$$

$$r_{x_ix_i}(\tau) = \frac{1}{T}\int_0^T r_{\tilde{x}_i\tilde{x}_i}(t+\tau, t) = \sum_{m} \frac{r_{x_ix_i}(m)}{T}(T-|\tau-mT|)$$

$$|\tau - mT| < T \tag{4.5}$$

where $r_{x_ix_i}(m)$ is as specified in (4.2). Thus $r_{x_ix_i}(m)$ is exponential whereas $r_{x_ix_i}(\tau)$ is piece-wise linear. These first and second order statistics of $x_i(t)$ are used in the estimation and prediction processes as described in the following sections.

4.2.1 Estimating Current Traffic State

The intensity of the packets transmitted on the output channel will vary as the traffic input from each of the active sources varies according to (4.1). Let $\mathbf{c} = \{c(j), c(j\text{-}1), ..., c(j\text{-}J\text{+}1)\}$ be the data vector representing the cell counts at the multiplexer output during small consecutive observation windows of size w. The estimation of the current traffic state implies obtaining an estimate $\hat{x}(j)$ of the actual output transmission rate $x(j)$, given **c**. The impulse response of the least mean-square filter should be such that the least mean-square error between the desired multiplexer output $x(j)$ and the filter estimate $\hat{x}(j)$ is minimized. This is achieved by solving the following normal equation [7]

$$r_{cc}(j) \otimes f(j) = g_{xc}(j) \tag{4.6}$$

where r_{cc} is the autocovariance function of $c(j)$; g_{xc} is the crosscovariance between $x(j)$ and $c(j)$; and $f(j)$ is the filter response. According to (2.6), $g_{xc}(j)$ is the convolution of $r_{cc}(j)$ and $f(j)$. Solving the above normal equation for filter coefficients $\boldsymbol{f} = \{f(1), f(2), ..., f(J\text{-}1), f(J)\}$ requires determining r_{cc} and g_{xc}, which is done as follows:

Considering $x_i(j)$ as the discrete form of $x_i(t)$ sampled at intervals w with $w \ll T$ and $wW = T$ the autocovariance function $r_{x_i x_i}(\tau)$ of (4.5) can be rewritten as

$$r_{x_i x_i}(j) = \sum_m \frac{r_{x_i x_i}(m)}{W}(W - |j - mW|) \qquad |j - mW| < W \tag{4.7}$$

Once again, for clarification purposes, n and m represent sampling at intervals T whereas j represents sampling at intervals w. As each of the active VBR source can be modeled as $x_i(t)$, the output process $x(j)$

which is an aggregation of s number of these independent and identical processes thus has an autocovariance function

$$r_{xx}(j) = s \times r_{x_i x_i}(j) \tag{4.8}$$

Now the unknowns r_{cc} and g_{xc} need to be determined in terms of the known r_{xx}, which depends on the statistical characteristics of the intra-frame arrival process. Firstly, a simple deterministic intra-frame arrival is assumed and thereafter the analysis is extended to the Poisson arrival mode.

4.2.1.1 Deterministic Arrival

If the intra-frame arrival is assumed to be deterministic then the relationships between the first and the second order statistics of processes $c(j)$ and $x(j)$ are given as follows

$$E\{c(j)\} = E\{E\{c(j)|x(j)\}\} = wE\{x(j)\} \tag{4.9}$$

and similarly

$$r_{cc}(j) = \begin{cases} w^2 r_{xx}(j) + \sigma_n^2 & j = 0 \\ w^2 r_{xx}(j) & j \neq 0 \end{cases} \tag{4.10}$$

$$g_{xc}(j) = w r_{xx}(j) \tag{4.11}$$

The term σ_n^2 reflects the noise that accounts for the measurement error and other random deviations from the model. Using (4.9), (4.10), (4.11), equation (4.6) can be rewritten in the matrix form as follows:

$$(w^2 \boldsymbol{R}_{xx} + \sigma_n^2 \boldsymbol{I})\boldsymbol{f} = w\boldsymbol{r}_{xx} \tag{4.12}$$

where $\boldsymbol{R}_{xx}$ is the $J \times J$ toeplitz autocovariance matrix of $x(j)$, and $\boldsymbol{f}$ and $\boldsymbol{r}_{xx}$ are $J \times 1$ column vectors. The causal finite impulse response of the least mean-square filter is obtained by solving (4.12) for J unknown elements of column vector f using matrix inversion *i.e.*,

$$\boldsymbol{f} = (w^2 \boldsymbol{R}_{xx} + \sigma_n^2 \boldsymbol{I})^{-1} w \boldsymbol{r}_{xx}. \tag{4.13}$$

4.2.1.2 Poisson Arrival

If the intra-frame arrival is assumed to be a Poisson distributed then [35]

$$p(c(j) = c | x(j) = x) = \frac{(xw)^c e^{-xw}}{c!}. \tag{4.14}$$

The relationships between the first and the second order statistics of processes $c(j)$ and $x(j)$ are then given as follows.

$$E\{c(j)\} = E\{E\{c(j)|x(j)\}\} = wE\{x(j)\} \tag{4.15}$$

$$r_{cc}(j) = \begin{cases} w^2 r_{xx}(j) + wE\{x(j)\} & j = 0 \\ w^2 r_{xx}(j) & j \neq 0 \end{cases} \tag{4.16}$$

$$g_{xc}(j) = w r_{xx}(j) \tag{4.17}$$

The term $wE\{x(j)\}$ accounts for the effects of burstiness over smaller observation windows, due to Poisson distribution of intra-frame cell arrival. Again, the causal filter response can be obtained as in (4.13).

Although a simple first order AR process is assumed in (4.1), it can be noticed that the order of the AR process impacts the above analysis only through the autocorrelation function. Thus, if a higher order AR process is needed to model the VBR video traffic, perhaps for accuracy reasons, the coefficients of the filter are still derived using a similar analysis. The main difference is that the autocorrelation function $r_{x_i x_i}(m)$, defined for an AR(1) process in (4.2), is replaced with that of an AR(I) process. A discussion on the derivation of the autocorrelation function of an AR(I) process is given in [7].

The next step is to predict the future traffic states.

4.2.2 Predicting Future Traffic States

It can be easily deduced from the discussion of the last section that since the multiplexer output is an aggregation of s independent and identical AR(1) modulated arrival processes, $\hat{x}(j)$ evolves according to an AR(1) process over a frame interval.

To be more specific

$$\hat{x}(j+W) = \alpha\hat{x}(j) + \beta\sum_{i=1}^{s} w_i(j+W). \tag{4.18}$$

With the autocovariance of $\hat{x}(j)$ as given in (4.8), prediction of $\hat{x}(j)$ over at least a few frame intervals is feasible. Again, a single element least mean-square filter, set up as a linear predictor, is used. The gain of this predictor is α^k for a k-step prediction. This is the best that could be done as the noise term in (4.18) is unpredictable and accounts for the prediction error. The prediction error, in general, is an i.i.d. Gaussian random variable with a variance (for a k-step prediction) formulated as

$$s\beta^2\sum_{l=1}^{k}\alpha^{2(k-l)}. \tag{4.19}$$

The variance of prediction error is thus proportional to s as well as k. If, on the other hand, an AR(I) process is assumed in (4.1) then the predictor will have I coefficients instead of one [7].

4.3 MPEG Sources

The broadcast quality MPEG sources are considered in this section. Although there is no particular model or class of models that can parsimoniously capture the characteristics of full range of MPEG sources, some interesting and relevant characterizations are presented by Doulamis *et al.* [13] and Lambardo *et al.* [27]. Doumalis *et al.* [13] decomposed MPEG traces into separate I, P and B frame series and, thereafter, modeled these series separately. These separate series were determined to be concatenation of stationary AR (AutoRegressive) processes, with parameters and, perhaps, the order changing at the scene boundaries. Furthermore, Lambardo *et al.* [27] suggested that the crosscorrelation among I, P and B frame series are time invariant.

It can, therefore, be inferred that the MPEG traffic from a single source, and consequently, the aggregate MPEG traffic is a concatenation of wide-sense cyclostationary processes that are also periodically AutoRegressive. In other words, if $c(j)$ be the packet count process at the multiplexer output, obtained by counting the number of packets transmitted over small consecutive observation windows of size w, then

$$c(j) = \sum_{i=1}^{N} \alpha_i(j-i)c(j-i) + e(j) \tag{4.20}$$

where N is the order of the PAR process; $\alpha_i(.)$ are the coefficients which are periodic with a period of a GOP interval denoted by T_{GOP}; and $e(.)$ is the white Gaussian noise. The mean and the autocorrelation function of the packet count process are also periodic *i.e.*,

$$E\{c(j)\} = E\{c(j+W)\} \tag{4.21}$$

and

$$r_{cc}(j) = r_{cc}(j+W) \tag{4.22}$$

where W is the number of observation windows within a GOP interval and, thus, reflects the sampling of the cell count process. Also, the observation window size $w << T_{GOP}$ and $wW = T_{GOP}$ This traffic structure is ideal for a time sequenced or periodic predictor, which is described next.

4.3.1 Time-Sequenced Adaptive Filters

Prediction of future traffic means determining $c(j+k)$, given the data vector $\boldsymbol{c} = \{c(j), c(j-1), ..., c(j-J+1)\}$. The order as well as the parameters of the PAR cyclostationary process, defined in (4.20), are usually unknown and, in fact, may dramatically change whenever there are scene changes in the underlying sequences. This calls for an adaptive algorithm that converges quickly. RLS algorithm is deemed to be the most suitable candidate for this application [7]. RLS algorithm orthogonalizes the input and is, therefore, less sensitive to the eigenvalue spread of the autocorrelation matrix of the input. Consequently, this algorithm is known to converge faster than NLMS (Normalized Least Mean Square) adaptive algorithm and its other variants that are used in [1], [8], [21], and [52], while still comparable as far as computational complexity is concerned. Some capability to track non-stationarity is also achieved in RLS algorithm, by introducing a parameter called forgetting factor and setting it to less than unity, based on the length of signal stationarity [7].

Since the cell count process has periodically recurring statistical properties, a time sequenced RLS predictor is used [16, 30]. The structure of this predictor is shown in Figure 4-2. It contains W sets of adaptable weights. Each set of weights is updated and, subsequently, used for producing output, periodically with a period of a GOP interval. This procedure results in a global optimal predictor that yields better performance in terms of least mean-square error for cyclostationary signals, as compared to a conventional adaptive filter, after convergence.

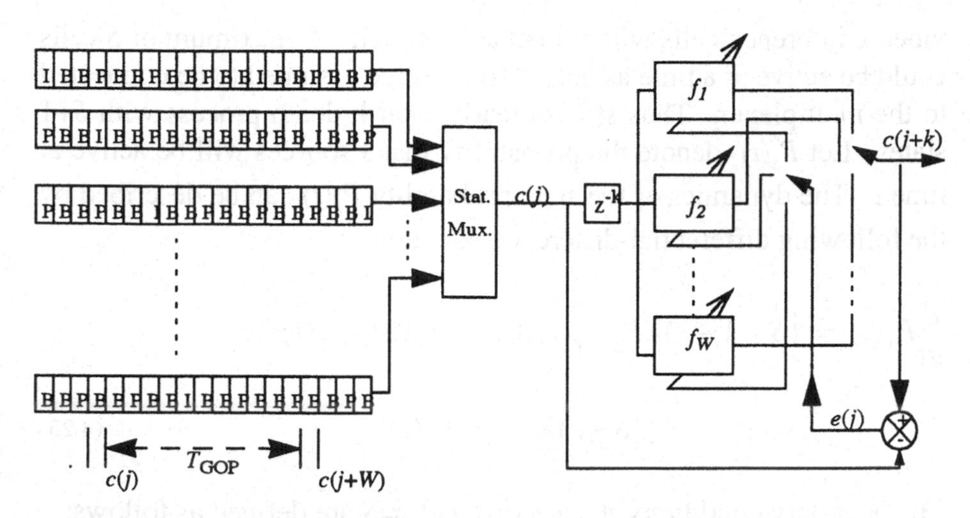

Figure. 4-2. Input Traffic and Time Sequenced Adaptive Predictor Structure.

4.4 Predicting Time of Congestion

The multiplexer is considered overloaded if $x(j)$ approaches or exceeds the total link capacity. Predicting the time of congestion, therefore, implies predicting the time at which the multiplexer will become overloaded, given the current state of the traffic. This is a two fold process that involves (i) predicting traffic in the active calls *i.e.*, predicting $\ddot{x}(j+k)$ given $\hat{x}(j)$; and (ii) estimating probability of arrival of new calls *i.e.*, predicting $s(t+\tau)$ given $s(t) = s$, where $s(t)$ is the number of VBR sources active at time t. Thus, if the predicted rate of transmission (resulting from all the active calls) at the output channel is approaching the channel capacity and the probability of new arrivals is also high, then the packet level controller must take suitable actions to avoid the onset of congestion.

The arrival of calls to the multiplexer is modeled as a Poisson process with average rate of arrival being λ. The holding time of each call is exponentially distributed with mean $1/\mu$ and is independent of earlier arrivals. The assumptions of Poisson call arrival and exponential call holding time are made to keep the analysis tractable. Besides, it can be argued that since the telephone and the tele-conference calls exhibit such characteristics, there is no reason why the video-phone and the video-conference calls will not behave as such. A maximum of S calls could be active at a time as only S sources are assumed to be connected to the multiplexer. Thus $s(t)$ is clearly a birth death process with $S+1$ states. Let $P_s(t)$ denote the probability that s sources will be active at time t. The dynamics of the state probability $P_s(t)$ can be described by the following differential-difference equation:

$$\frac{d}{dt}P_s(t) = (S+1-s)\lambda P_{s-1}(t) + (s+1)\mu P_{s+1}(t)$$

$$-[(S-s)\lambda + s\mu]P_s(t). \tag{4.23}$$

The boundary conditions at state $s=0$ and $s=S$ are defined as follows:

$$\frac{d}{dt}P_0(t) = \mu P_1(t) - S\lambda P_0(t), \tag{4.24}$$

$$\frac{d}{dt}P_S(t) = \lambda P_{S-1}(t) - S\mu P_S(t). \tag{4.25}$$

Let $P(z,t)$ be the probability generating function; and considering an arbitrary initial condition that at time $t=0$, i sources were active *i.e.*, $P_i(0) = 1$ when $0 \le i \le S$ and $P_s(0) = 0$ when $s \ne i$, we get (Appendix A)

$$P(z,t) = \left(\frac{z+\frac{\mu}{\lambda}}{G+\frac{\mu}{\lambda}}\right)^S G^i \quad \text{where} \quad G = \frac{1+\frac{\mu}{\lambda}\left(\frac{z-1}{z+\frac{\mu}{\lambda}}\right)e^{-(\mu+\lambda)t}}{1-\left(\frac{z-1}{z+\frac{\mu}{\lambda}}\right)e^{-(\mu+\lambda)t}}. \tag{4.26}$$

The state probabilities $P_S(t)$ can now be obtained by expanding the probability generating function in a power series. The coefficients of this power series then are the state probabilities $P_S(t)$ *i.e.*,

$$P_s(t) = \frac{1}{s!}\frac{\partial^s}{\partial z^s}P(z, t)\bigg|_{z = 0} \tag{4.27}$$

which is the probability that s sources will be active at time t, given that initially there were i sources active. Thus if $s = c$ be the state of the process $s(t)$, such that if c number of sources become active at the same time it will cause the multiplexer to overload, the distribution $P_c(t)$ can be obtained by using the above expression to estimate the time to overload, which is the time when the process $s(t)$ is expected to pass through the state $s = c$ for the first time. The parameter c is the maximum number of allowed VBR video connections in a VP, and it is determined based on the desired level of service availability as well as the revenue considerations. A procedure that iteratively determines optimal values of such call admission control parameters for each link in the network is outlined in the later chapters.

4.5 Results

The performance of the proposed technique for estimating and predicting multiplexed traffic is evaluated in this section. Traffic traces for video-conferencing and video-phone sources are generated using the autoregressive model of (4.1), where the values of α and β are taken to be 0.8781 and 0.1108 respectively, and $w(n)$ as iid Gaussian white noise with mean = 0.572 and variance = 1 as specified in [31]. This produces traffic with mean transmission rate of 3.9 Mbits/sec, peak transmission rate of 10.575 Mbits/sec and standard deviation of 1.725 Mbits/sec. Cases of up to S=10 (the maximum number of video sources) are implemented.

First, the intra-frame arrival process is simulated as being deterministic. Figure 4-3(a) illustrates the performance of least mean-square filter as compared to the direct estimation of $x(j)$ *i.e.*, $c(j)/w$ in the presence of noise. Up to 50 trials are conducted for each increment of noise to signal power *i.e.*, $\sigma_n^2/r_{cc}(0)$. In each trial up to 10 VBR sources are activated with their phases uniformly distributed between $[0,T]$ where T is 1/30 sec. The ratio of the error variance resulting from direct estimate to the least mean-square error resulting from least mean-square filter estimate, is obtained at each trial and averaged over 50 trials. Figure 4-3(a) shows that the least mean-square filter estimates, in the presence of noise, are always better than the direct estimate of $x(j)$; and this relative performance becomes more profound as the noise power in the signal increases.

Next, similar set of trials are conducted for the cases where the intra-frame arrival process is simulated as being Poisson. In Figure 4-3(b) the ratio of the error variance resulting from direct estimate of $x(j)$ to the least mean-square error resulting from least mean-square filter estimate is plotted as a function of observation window size. Figure 4-3(b) shows that the least mean-square filter estimates are always better than the direct estimate of $x(j)$; and again this relative performance becomes more profound as the observation window size decreases. The observations of Figure 4-3 also validate the model of section 4.2.

Once the task of estimating the current activity state of the sources is accomplished, the next step is to determine the time to overload. Figure 4-4 shows a typical sample function of the process $x(j)$ obtained by aggregating 10 VBR sources with their phases uniformly distributed between $[0,T]$. Figure 4-4 also shows some examples of linear prediction of such sample functions. The d.c. component or the mean is removed to eliminate the unnecessary bias in the prediction error. These results show that the predictor tracks $x(j)$ fairly accurately as long as the prediction is over a few frame intervals. However, as expected in (4.19), the prediction error becomes more prominent as the prediction step size increases.

The performance of time-sequenced adaptive filters is evaluated using a number of half an hour long empirical traces of broadcast quality video as input traffic [44]. Multiple traces are aggregated and the resulting traffic is predicted. Both, NLMS as well as RLS adaptive algorithms, are employed for the purpose of a comparative analysis. For simplicity, the number of sets of weights *i.e.*, W are chosen to be 12, based on the GOP size of the sequences, which is also 12.

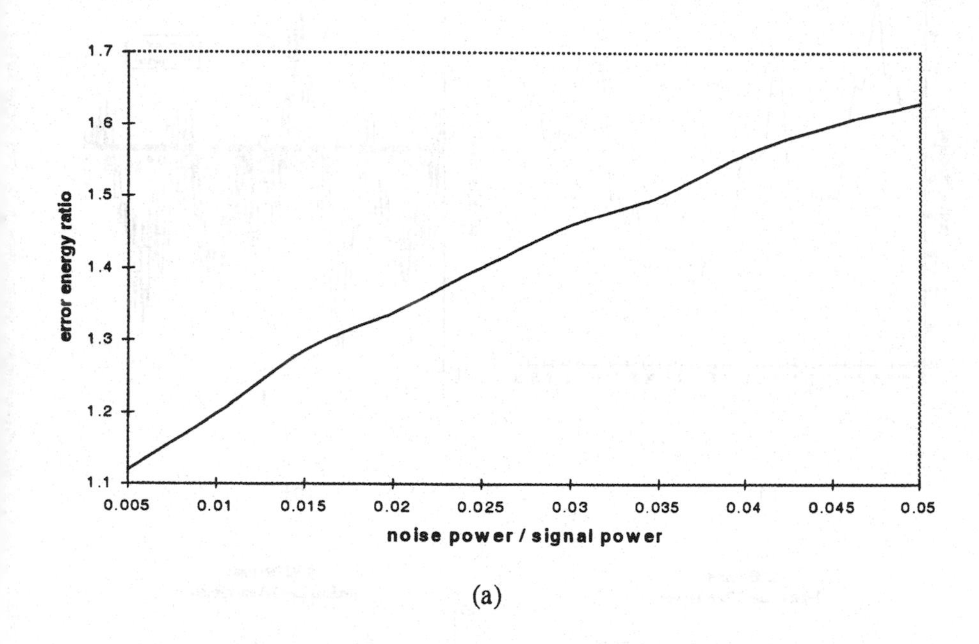

(a)

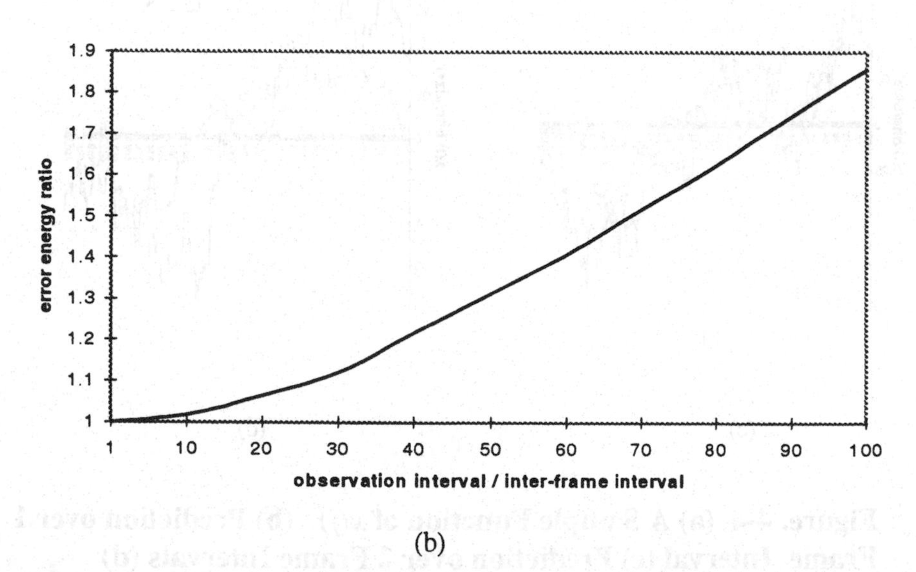

(b)

Figure. 4-3. Estimation Performance of Least Mean-Square Filter (a) Deterministic Arrival (b) Poisson Arrival.

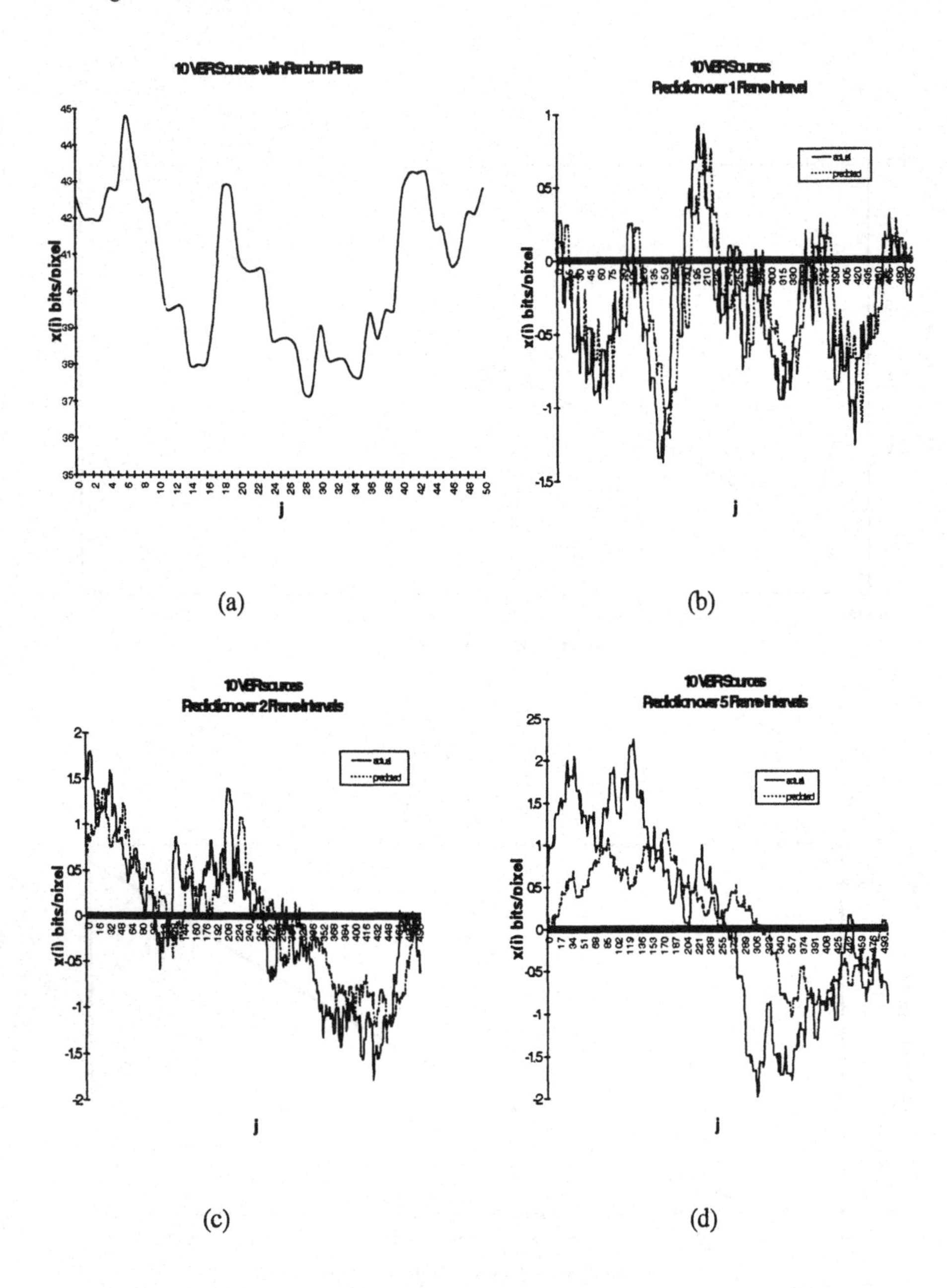

Figure. 4-4. (a) A Sample Function of $x(j)$ (b) Prediction over 1 Frame Interval (c) Prediction over 2 Frame Intervals (d) Prediction over 5 Frame Intervals.

Table 4-1 shows the performance comparison between NLMS and RLS, when used in the filter structure of Figure 4-2, for predicting a single source. The metric used for comparison is $SNR^{-1} = \frac{\sum e^2(n)}{\sum c_i^2(n)}$, where the numerator is the steady-state mean-square prediction error, and the denominator is the signal power. Results in Table 4-1 show that the RLS algorithm clearly outperforms NLMS algorithm in terms of better signal to noise ratio because of its faster convergence, subsequent to a scene change.

Figure 4-5 shows some samples of the actual multiplexed traffic and the corresponding predicted traffic. The starting times of the traces are chosen at random. The predictor tracks the traffic quite closely for most of the times except when there are random transitions in the underlying traffic. It is quite difficult to characterize or formulate the prediction error, as the exact parameters of the underlying traffic are unknown. Some trends, however, can be noticed from Figure 4-5 such as, adding more sources adds to the prediction error. This happens because the sources are uncorrelated and, therefore, the prediction error is additive. The signal-to-noise ratio, however, would stay proportional to the values presented in Table 4-1, or would degrade gracefully; because adding more sources adds to the signal power as well.

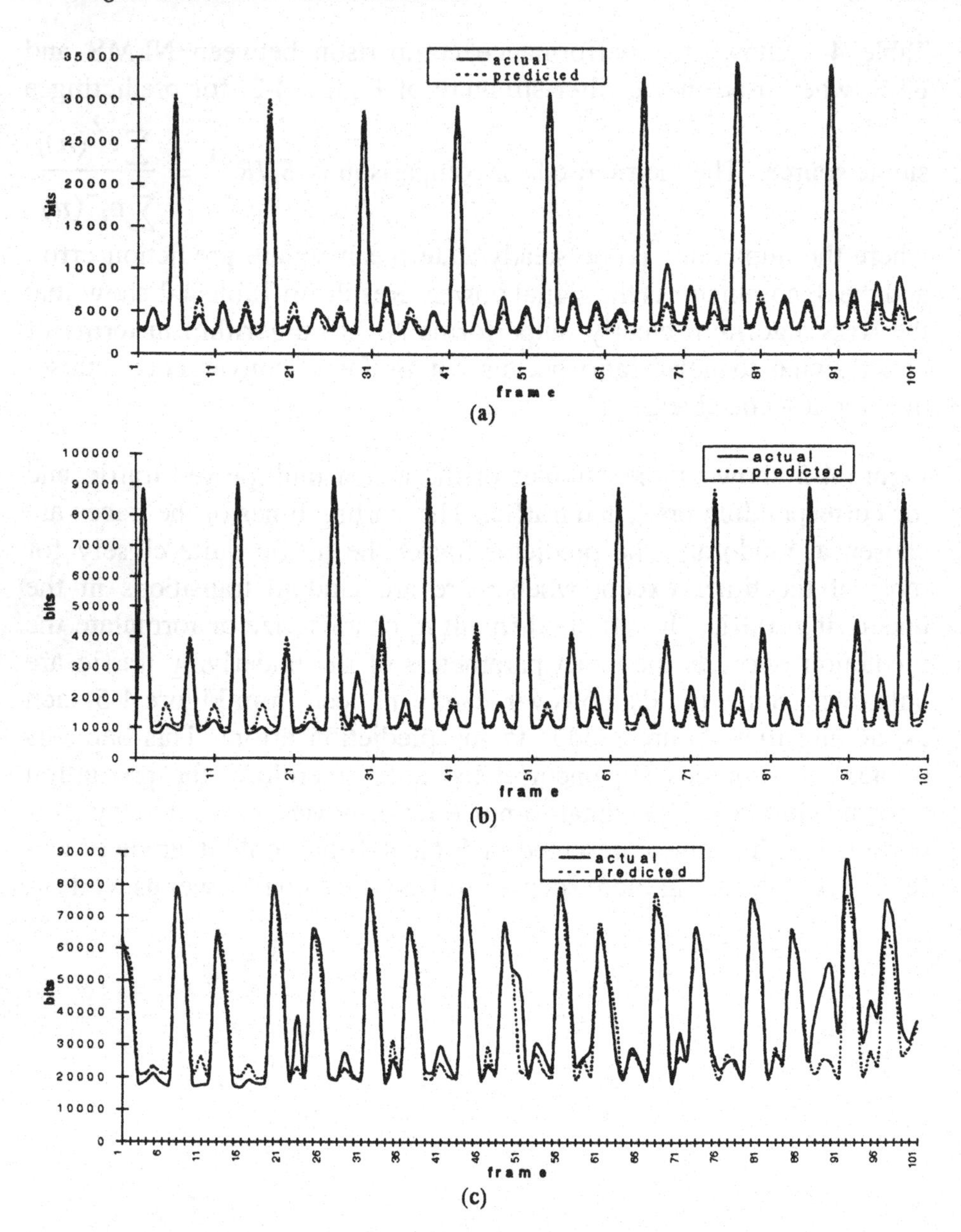

Figure. 4-5. Traffic Prediction (a) Single Source (Terminator Trace) (b) Three Sources Multiplexed with Random Starting Times (c) Five Sources Multiplexed with Random Starting Time.

TABLE 4-1. Relative Performance of NLMS and RLS. The metric used is SNR^{-1}. [x,y] means the range.

Sequence	Weight Set Index	NLMS	RLS
Terminator	1	.026	.0244
	2,3,5,6,8,9,11,12	[.224, .242]	[.212, .229]
	4,7,10	[.176, .183]	[.170, .177]
Star Wars	1	.0143	.0140
	2,3,5,6,8,9,11,12	[.23, .26]	[.225, .232]
	4,7,10	[.312, .338]	[.278, .290]
Talk Show	1	.0047	.0046
	2,3,5,6,8,9,11,12	[.023, .06]	[.02, .09]
	4,7,10	[.123, .146]	[.08, .11]
Dino	1	.0122	.011
	2,3,5,6,8,9,11,12	[.129, .158]	[.11, .16]
	4,7,10	[.173, .178]	[.161, .167]
Silence of the Lambs	1	.0206	.020
	2,3,5,6,8,9,11,12	[.156, .189]	[.14, .18]
	4,7,10	[.293, .363]	[.25, .32]
Bond	1	.0171	.0170
	2,3,5,6,8,9,11,12	[.051, .06]	[.051, .079]
	4,7,10	[.073, .081]	[.071, .079]
Asterix	1	.026	.0251
	2,3,5,6,8,9,11,12	[.181, .199]	[.170, .190]
	4,7,10	[.16, .178]	[.160, .169]
Simpsons	1	.020	.0197
	2,3,5,6,8,9,11,12	[.234, .362]	[.221, .482]
	4,7,10	[.294, .312]	[.243, .275]

The B-ISDN networks are expected to have a large delay-bandwidth product. In these networks, if a destination node becomes congested and sends a control packet to the source of the problem, a large number of packets may have already been launched into the network, causing further deterioration, before any control action takes effect. For example, if the round-trip propagation path is 1000 km long then, with a propagation delay of 5μsec/km along that path (this value is fairly typical for high-speed fiber), the round-trip propagation delay is 5msec [45]. During this interval, assuming a transmission capacity of 45 Mbps, about 500 ATM cells could have already been enroute. The results in Figure 4-5 demonstrate a 12-frame prediction which, assuming a frame rate of 30 Frames/sec, implies capability to predict traffic about 400 msec ahead of time. Considering the aforesaid delay-bandwidth characteristic of B-ISDN, this is a significant enhancement to the congestion avoidance capabilities.

Figure 4-6 illustrates a sample of a situation where three sources, with random starting times, are being multiplexed. One of the sources gets deactivated and after 50 GOP intervals, a new source gets activated. The output from only one set of weights is shown for the clarity of the figure. Again, it can be observed that the RLS algorithm does a better job of tracking such sharp transitions in the input traffic.

Lastly, Figure 4-7 depicts the probability distribution to overload, computed using (4.27). The average active time ($1/\mu$) and the average arrival rate λ are arbitrarily taken to be 10 sec and 1/15 sec respectively, with μ/λ then being 1.5. Assuming arbitrarily that c=7, the plots in Figure 4-7 represent the transient state probability distribution $P_7(t)$ given that at time t=0, i (the number of sources initially active) is 0, 1, 2, & 3 respectively. For example, given that initially there are three active VBR calls, the probability that four more calls will arrive within 10 sec, from Figure 4-7, is about 0.03. Since an analytical solution to (4.27) is extremely difficult to obtain, numerical methods approach (using Maple software) is used to obtain the transient state probability distribution. It can also be noticed that as $t \to \infty$, $P_s(t) \to P_s$ *i.e.*, it reaches steady state, which is 0.04246 for this case. Figure 4-7 clearly shows that under the assumptions of the birth-death characteristics of the process $s(t)$, the time to overload becomes problematic only when the initial state i is already close to c.

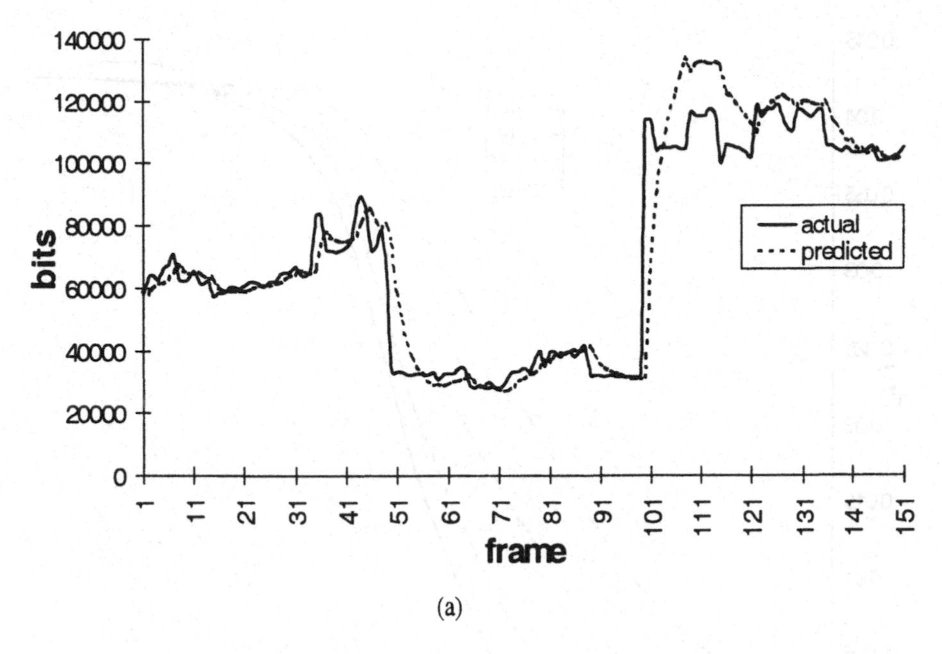

(a)

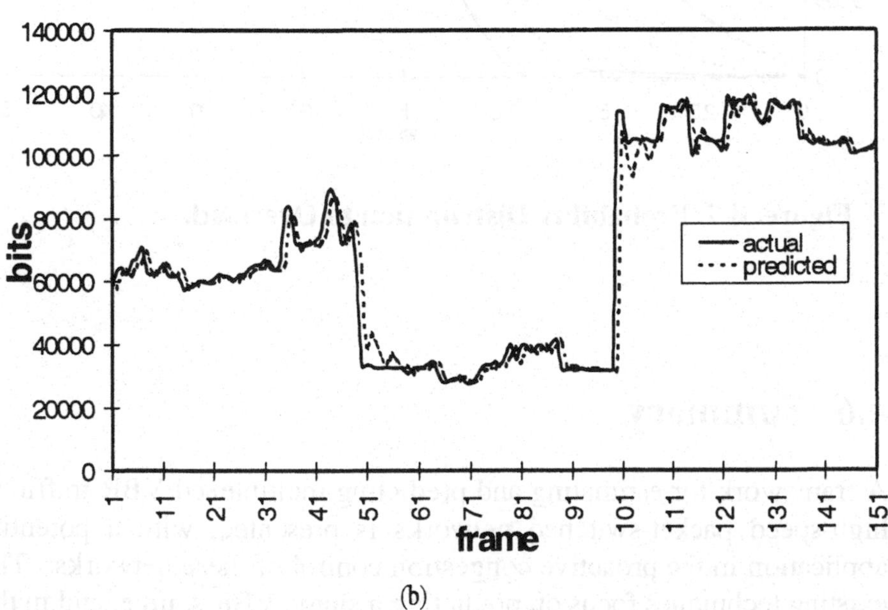

(b)

Figure. 4-6. Transient Behavior in Response to Sources Turning Off and On (a) using LMS (b) using RLS.

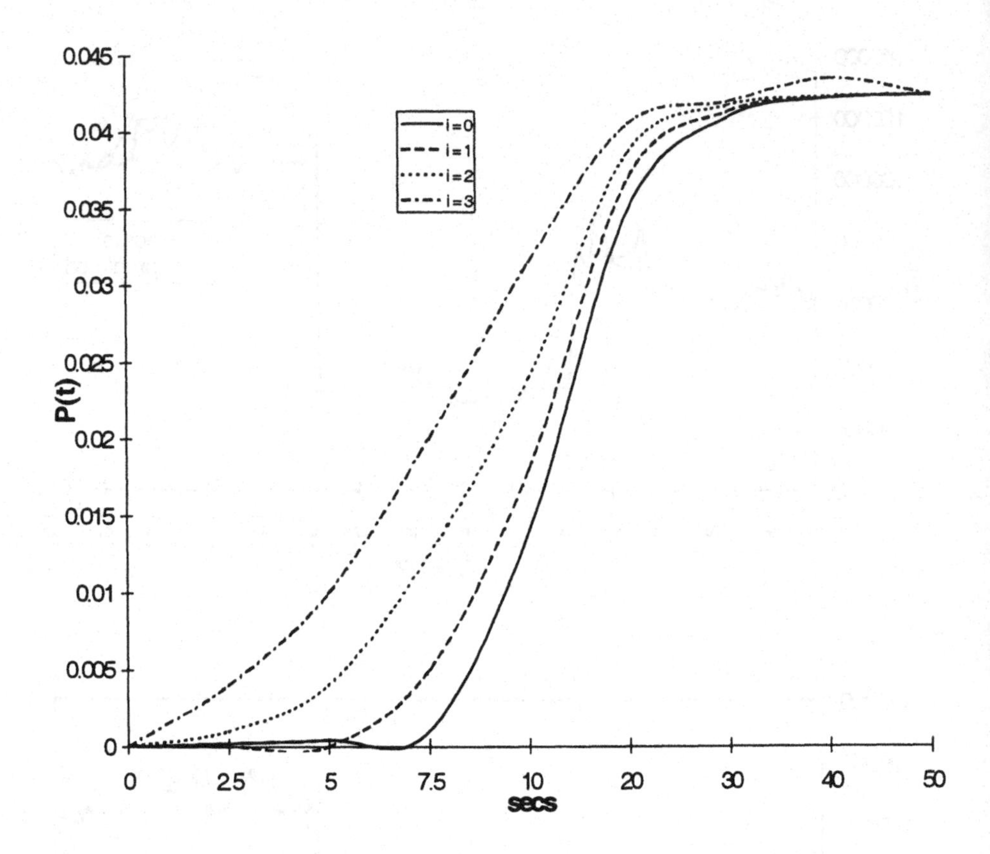

Figure. 4-7. Probability Distribution to Overload.

4.6 Summary

A framework for estimating and predicting multiplexed VBR traffic in high-speed packet-switched networks is presented, with a potential application in the proactive congestion control of these networks. The existing techniques focus on predicting a single VBR source, and in the case of MPEG sources the traffic is predicted by first decomposing the source traffic into I, P and B frame series and then forecasting each series separately. These techniques, thus, can be effectively used only in the proximity of the source. The proposed technique does not have such restrictions and could be used, ad-hoc, at any VP in the network which is transporting homogenous multiplexed MPEG traffic. The number of packets being transmitted on the VP are counted and the future traffic is forecasted by exploiting the cyclostationary and autoregressive nature of the multiplexed VBR traffic. Time invariant linear filters are proposed for situations where the source traffic model parameters are known. Time sequenced adaptive filters are proposed when the model parameters are unknown. Simulation results demonstrate the practicability of the approach. A future extension of this work would be to test the proposed technique in the presence of background voice and low-priority data traffic.

Next, some priority-based channel access control policies are proposed and evaluated. As suggested in Chapter 1, for a packet-switched network, with VBR traffic, the call admission control alone is not sufficient to ensure a congestion-free flow of traffic, unless the network is under-provisioned. However, a larger number of VBR calls can be admitted, without violating the packet-level QoS guarantees, provided the network is capable of predicting the onset of congestion, and initiating a suitable control action, ahead of time. The estimation and prediction based approach to congestion avoidance, proposed in this chapter, is thus complementary to the call admission control discussed in the next chapter.

Chapter 5

Channel Access Control

In this chapter the performance of various channel access control policies is evaluated in terms of call level QoS parameters such as call blocking probability and call dropping probability, using multi-dimensional Markov chain analysis. First some basic conventional policies are discussed to lay proper ground work, and thereafter the two priority based policies are analyzed.

A wide spectrum of bandwidth access control policies have been discussed in the literature [2, 36]. The simplest of all policies is a complete sharing (CS) policy, which permits an unrestricted sharing of total bandwidth among all the competing traffic types. On the other extreme is a complete partitioning (CP) policy, which permanently and statically partitions the bandwidth among the competing services. It is evident that the most desirable policy would lie between these two extremes. Partial sharing (PS) is one such class of schemes [2, 36]. According to the partial sharing policy, the total capacity can be partitioned into engineered capacity and shared capacity. The engineered capacity is intended to reserve proper capacity for accommodating the expected call arrivals, whereas the shared capacity is used in reducing the impact of fluctuations in the arrival rate. Determination of accurate sizes of shared and engineering capacities, for every possible network configuration and load conditions, particularly when these conditions are dynamically varying, could become quite complex. Two priority based schemes are, therefore, proposed namely 'PS with Call Dropping' and 'PS with Discouraged Arrivals'. It is demonstrated using numerical results, in this chapter, that the priority based policies significantly improve QoS for high priority services at the expense of low priority services.

Although the priority-based policies involving preemptive priority or discouraged arrivals have been discussed previously in literature [5], they have mostly been studied under a multi-service single-server queue regime. The main contribution of this chapter is that these policies have been thoroughly analyzed for the first time for a loss-loss, as well as, a mixed loss-delay multi-service multi-server system, where services have heterogeneous bandwidth requirements.

5.1 Call Model

Consider a link of capacity C supporting traffic from I services. The traffic from service s_i arrives at a Poisson rate λ_i with exponentially distributed holding time of mean $1/\mu_i$. Each call from s_i occupies m_i units of capacity for the duration of the call. For VBR sources m_i may be the peak rate or the equivalent bandwidth depending upon the packet level QoS guarantees. This is an I-dimensional Birth-Death Markov process, with vector $\boldsymbol{j} = \{j_1, j_2, \ldots, j_i, \ldots, j_I\}$ representing the state of the link *i.e.*, the number of connections from each service that could be active. The link state j depends on the channel access control policy implemented for that link. The objective now is to determine the probability of being in state $\boldsymbol{j}$ *i.e.*, $P(j)$ and, then compute the blocking probability Pb_i which is the probability that a call from service s_i will be blocked at this link. Assuming that the call arrival process is stationary, $P(j)$ is obtained by solving the steady-state equilibrium equations of the I-dimensional Birth-Death Markov process. For simplicity a two service case is assumed *i.e.*, $I = 2$. The analysis could be easily extended to a multi-service case.

5.2 CS (Complete Sharing)

As mentioned earlier, complete sharing implies uncontrolled access to the link's bandwidth by any of the services. Thus, as long as $m_1(m_2)$ units of bandwidth are available, a call from $s_1(s_2)$ is accepted. All the blocked calls are assumed to be lost. The acceptable states, as well as, state transitions are shown in Figure 5-1(a). The rate of transition from (j_1,j_2) to (j_1+1,j_2) is λ_1; from (j_1,j_2) to (j_1,j_2+1) it is λ_2; from (j_1,j_2) to (j_1-1,j_2) it is $j_1\mu_1$; and from (j_1,j_2) to (j_1,j_2-1) it is $j_2\mu_2$. In a multi-service case the acceptable states will be in the region bounded by a finite number of hyper-planes. At equilibrium, the steady state global balance equations are [2]

$$[\lambda_1\delta_{j_1+1,j_2} + \lambda_2\delta_{j_1,j_2+1} + j_1\mu_1\delta_{j_1-1,j_2} + j_2\mu_2\delta_{j_1,j_2-1}]P(j_1,j_2)$$

$$= \lambda_1\delta_{j_1-1,j_2}P(j_1-1,j_2) + \lambda_2\delta_{j_1,j_2-1}P(j_1,j_2-1) +$$

$$(j_1+1)\mu_1\delta_{j_1+1,j_2}P(j_1+1,j_2) + (j_2+1)\mu_2\delta_{j_1,j_2+1}P(j_1,j_2+1) \quad (5.1)$$

$$\sum_{(j_1,j_2)\in A} P(j_1,j_2) = 1 \quad (5.2)$$

for all $(j_1,j_2) \in A$. A is the space of acceptable states, which for this case is given as

$$A = \{(j_1,j_2) : j_1m_1 + j_2m_2 \le C\} \quad (5.3)$$

and

$$\delta_{k_1,k_2} = \begin{cases} 1 & if \quad (k_1,k_2) \in A \\ 0 & if \quad (k_1,k_2) \notin A \end{cases} . \quad (5.4)$$

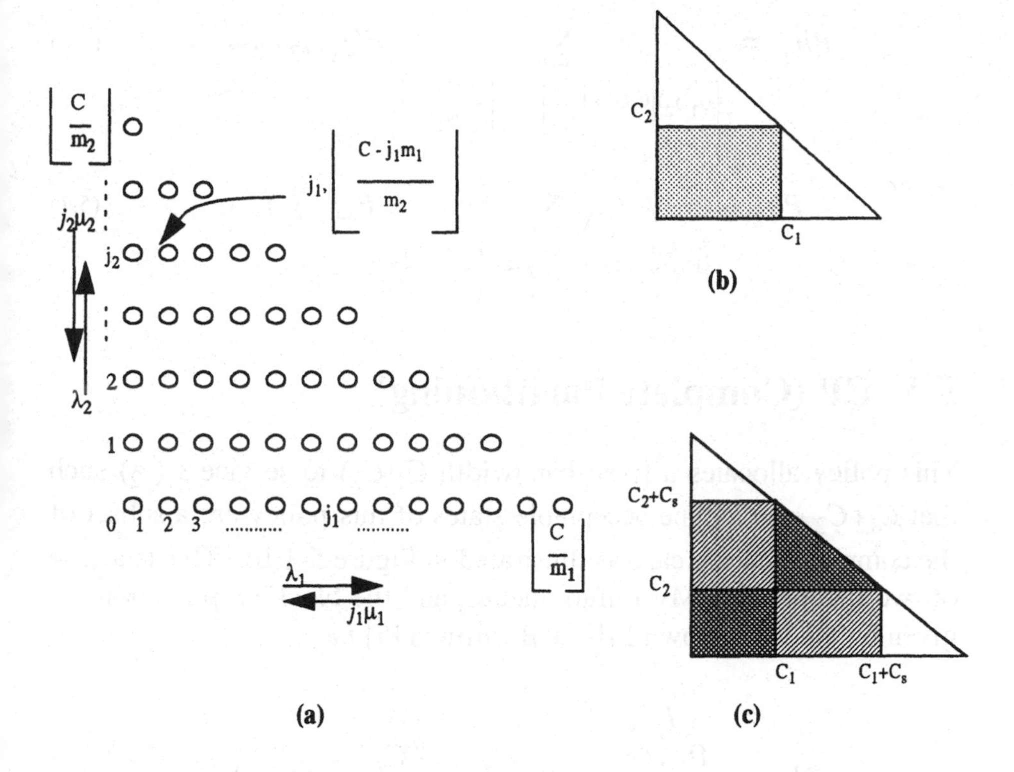

Figure. 5-1. Two Dimensional Markov chain for (a) Complete Sharing (b) Complete Partitioning (c) Partial Sharing.

Solving (5.1) and (5.2) we get the product-form solution [5] *i.e.*,

$$P(j_1, j_2) = (P_A(0))^{-1} \times \frac{\rho_1^{j_1}}{j_1!} \times \frac{\rho_2^{j_2}}{j_2!} \tag{5.5}$$

where $P_A(0) = \sum_{(j_1, j_2) \in A} \frac{\rho_1^{j_1}}{j_1!} \times \frac{\rho_2^{j_2}}{j_2!}$; and ρ_i for $i = 1, 2$ is the traffic intensity *i.e.*, λ_i/μ_i. The blocking probability is then the probability that the system will move out of the acceptable space A, and is given as

$$Pb_1 = \sum_{\left\{(j_1,j_2)\mid 0 \le j_2 \le \left\lfloor \frac{C}{m_2} \right\rfloor, j_1 = \left\lfloor \frac{C - j_2 m_2}{m_1} \right\rfloor\right\}} P(j_1, j_2) \text{ and} \tag{5.6}$$

$$Pb_2 = \sum_{\left\{(j_1,j_2)\mid 0 \le j_1 \le \left\lfloor \frac{C}{m_1} \right\rfloor, j_2 = \left\lfloor \frac{C - j_1 m_1}{m_2} \right\rfloor\right\}} P(j_1, j_2). \tag{5.7}$$

5.3 CP (Complete Partitioning)

This policy allocates a fixed bandwidth $C_1(C_2)$ to service $s_1(s_2)$ such that $C_1+C_2 <= C$. The acceptable states of this policy are a subset of the complete sharing case as illustrated in Figure 5-1(b). This is a case of two independent M\M\m\m queues, and the blocking probability is given by the well known Erlang-B formula [5] *i.e.*,

$$Pb_i = \frac{\rho_i^{J_i} / J_i}{\sum_{j=0}^{J_i} \rho_i^j / j!} \text{ where } J_i = \left\lfloor \frac{C_i}{m_i} \right\rfloor \text{ and } i = 1,2. \tag{5.8}$$

5.4 PS (Partial Sharing)

In partial sharing, some bandwidth $C_1(C_2)$ is allocated permanently to service $s_1(s_2)$, whereas a portion C_s could be shared by the two on a first-come-first-served basis, such that $C_1+C_2+C_s <= C$. This is a special combination of the above two policies. The acceptable states are illustrated in Figure 5-1(c). The blocking probabilities are

$$Pb_1 = \sum_{\left\{(j_1,j_2)\mid 0 \le j_2 \le \left\lfloor \frac{C_s + C_2}{m_2} \right\rfloor, j_1 = min\left(\left\lfloor \frac{C_s + C_1}{m_1} \right\rfloor, \left\lfloor \frac{C - j_2 m_2}{m_1} \right\rfloor\right)\right\}} P(j_1, j_2) \tag{5.9}$$

$$Pb_2 = \sum_{\left\{(j_1,j_2)\,|\,0 \le j_1 \le \left\lfloor \frac{C_s + C_1}{m_1} \right\rfloor,\, j_2 = min\left(\left\lfloor \frac{C_s + C_2}{m_2} \right\rfloor, \left\lfloor \frac{C - j_1 m_1}{m_2} \right\rfloor\right)\right\}} P(j_1, j_2). \quad (5.10)$$

The two priority based schemes are analyzed next.

5.5 CD (PS with Call Dropping)

In this case a call from a low priority service is accepted if enough capacity is available, however, the call could be dropped if a higher priority call arrives. We assume that s_2 is high priority broadband service and s_1 is low priority narrow-band service. The acceptable states, as well as, the state transitions are shown in Figure 5-2(a). The rate of transition from (j_1,j_2) to (j_1+1,j_2) is λ_1; from (j_1,j_2) to (j_1,j_2+1) it is λ_2; from (j_1,j_2) to (j_1-1,j_2) it is $j_1\mu_1$; and from (j_1,j_2) to (j_1,j_2-1) it is $j_2\mu_2$. Besides, the transitions out of a state, where an s_1 call could be dropped to make room for an incoming s_2 call, occur at a rate of λ_2 (Appendix B). At equilibrium the steady state global balance equations are

$$[\lambda_1 \delta_{j_1+1, j_2} + \lambda_2 \delta_{j_1, j_2+1} + j_1 \mu_1 \delta_{j_1-1, j_2} + j_2 \mu_2 \delta_{j_1, j_2-1} +$$
$$\lambda_2 \Delta_{j_1, j_2+1}] P(j_1, j_2) = \lambda_1 \delta_{j_1-1, j_2} P(j_1 - 1, j_2) +$$
$$\lambda_2 \delta_{j_1, j_2-1} P(j_1, j_2 - 1) + (j_1 + 1)\mu_1 \delta_{j_1+1, j_2} P(j_1 + 1, j_2) +$$
$$(j_2 + 1)\mu_2 \delta_{j_1, j_2+1} P(j_1, j_2 + 1) + \lambda_2 \sum_i \Delta_{j_1+i, j_2-1} P(j_1 + i, j_2 - 1)$$

$$(5.11)$$

$$\sum_{(j_1,j_2) \in A} P(j_1, j_2) = 1 \quad (5.12)$$

$$\Delta_{k_1,k_2} = \begin{cases} 1 \;\; if \;\; 0 \le k_2 \le \left\lfloor \frac{C}{m_2} \right\rfloor \; and \; \left\lfloor \frac{C - k_2 m_2}{m_1} \right\rfloor - b + 1 \le k_1 \le \left\lfloor \frac{C - k_2 m_2}{m_1} \right\rfloor \\ 0 \;\; otherwise \end{cases} \tag{5.13}$$

for all $(j_1, j_2) \in A$, where A and δ_{k_1,k_2} are as defined above, and

$b = \left\lfloor \frac{m_2}{m_1} \right\rfloor$.

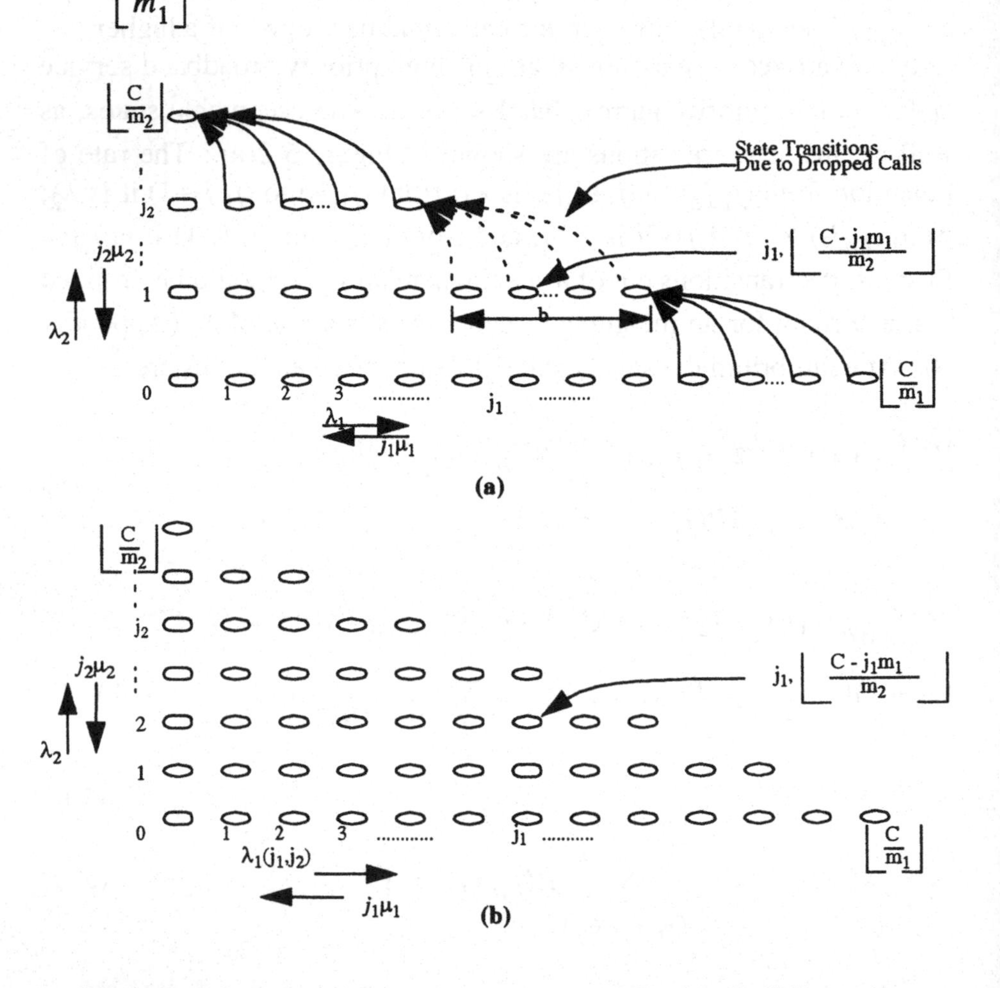

Figure. 5-2. Priority Based Policies (a) PS with Call Dropping (b) PS with Discouraged Arrivals.

Solving (5.11) and (5.12) we get the state probabilities $P(j_1, j_2)$. Since s_2 does not have to compete for bandwidth, its blocking probability is determined using the Erlang-B formula *i.e.*,

$$Pb_2 = \frac{\rho_2^{J_2}/J_2}{\sum_{j=0}^{J_2} \rho_2^{j}/j!} \quad \text{where } J_2 = \left\lfloor \frac{C_2}{m_2} \right\rfloor, \tag{5.14}$$

whereas the blocking probability for s_1 is, again, obtained by substituting $P(j_1, j_2)$ in (5.6). The call dropping probability is determined as follows:

It is clear from Figure 5-2(a) that an s_2 call may drop upto b active s_1 calls. The average s_1 calls that could be dropped are then $\sum_{i=1}^{b} iD_{s_i}$. D_{s_i} is the probability that the link is in the state where i number of s_1 calls would be dropped if an s_2 call arrives, and is given as

$$D_{s_i} = \sum_{\left\{(j_1, j_2) \mid 0 \le j_2 \le \left\lfloor \frac{C}{m_2} \right\rfloor - 1,\, j_1 = \left\lfloor \frac{C - j_2 m_2}{m_1} \right\rfloor - b + i \right\}} P(j_1, j_2) \quad \text{for } i = 1, 2, .. b. \tag{5.15}$$

Let D_s be the probability of being in any one of the states where a call could be dropped *i.e.*, $D_s = \sum_{i=1}^{b} D_{s_i}$. Let $x_1(x_2)$ be the random variable representing the number of $s_1(s_2)$ calls arriving in an interval T, and x_{1_d} be the random variable representing the number of s_1 calls dropped in the interval T, then

$$E\left\{x_{1_d}\Big|_{x_2 = a}\right\} = aD_s \sum_{i=1}^{b} iD_{s_i}. \tag{5.16}$$

$$E\{x_{1_d}\} = E\left\{E\left\{x_{1_d}\Big|_{x_2}\right\}\right\} = E\{x_2\}D_s \sum_{i=1}^{b} iD_{s_i} \tag{5.17}$$

and, therefore,

$$Pd_1 = \frac{E\{x_{1_d}\}}{E\{x_1\}} = \frac{\lambda_2 D_s \sum_{i=1}^{b} iD_{s_i}}{\lambda_1} \tag{5.18}$$

Dropping of calls, however, may not always be desirable. Two variants of this scheme which avoid call dropping by introducing call waiting are discussed below.

5.5.1 DQ (Dropped Calls Queued)

In this case the dropped calls stay in the system and continue to retry for channel access until serviced on a FIFO basis. The Markov chain process for this policy is shown in Figure 5-3(a). The rate of transition from (j_1,j_2) to (j_1+1,j_2) is λ_1 if $((j_1+1)m_1 + j_2m_2)$ is less than C, otherwise, it is zero. The rate of transition from (j_1,j_2) to (j_1-1,j_2) is $j_1\mu_1$ if $j_1 \le \left\lfloor \frac{C - j_2m_2}{m_1} \right\rfloor$, otherwise, it is $\left\lfloor \frac{C - j_2m_2}{m_1} \right\rfloor \mu_1$. From (j_1,j_2) to (j_1,j_2+1), and from (j_1,j_2) to (j_1,j_2-1), the rates of transition are λ_2 and $j_2\mu_2$, respectively. At equilibrium, the steady state global balance equations are:

$$[\lambda_1 \delta_{j_1+1, j_2} \Delta_{j_1, j_2} + \lambda_2 \delta_{j_1, j_2+1} + min\left(j_1, \left\lfloor \frac{C - j_2 m_2}{m_1} \right\rfloor\right) \mu_1 \delta_{j_1-1, j_2} +$$

$$j_2 \mu_2 \delta_{j_1, j_2-1}] P(j_1, j_2) \ = \lambda_1 \delta_{j_1-1, j_2} \Delta_{j_1-1, j_2} P(j_1-1, j_2) +$$

$$\lambda_2 \delta_{j_1, j_2-1} P(j_1, j_2-1) + min\left(j_1+1, \left\lfloor \frac{C - j_2 m_2}{m_1} \right\rfloor\right) \mu_1 \delta_{j_1+1, j_2} \bullet$$

$$P(j_1+1, j_2) + (j_2+1) \mu_2 \delta_{j_1, j_2+1} P(j_1, j_2+1) \tag{5.19}$$

$$\sum_{(j_1, j_2) \in A} P(j_1, j_2) \ = \ 1. \tag{5.20}$$

$$\Delta_{k_1, k_2} = \begin{cases} 1 & if \quad k_1 m_1 + k_2 m_2 < C \\ 0 & otherwise \end{cases} \tag{5.21}$$

for all $(j_1, j_2) \in A$. A is the space of acceptable states, which for this case is given as

$$A = \left\{ (j_1, j_2) : \left(0 \le j_1 \le \left\lfloor \frac{C}{m_1} \right\rfloor\right) and \left(0 \le j_2 \le \left\lfloor \frac{C}{m_2} \right\rfloor\right) \right\}, \tag{5.22}$$

and δ_{k_1, k_2} is as defined above. Pb_2 is obtained using (5.14), and Pb_1 is given as

$$Pb_1 = \sum_{\left\{ (j_1, j_2) \mid 0 \le j_2 \le \left\lfloor \frac{C}{m_2} \right\rfloor, \left\lfloor \frac{C - j_2 m_2}{m_1} \right\rfloor \le j_1 \le \left\lfloor \frac{C}{m_1} \right\rfloor \right\}} P(j_1, j_2). \tag{5.23}$$

The average number of s_1 calls in the system is given as

$$E(j_1) = \sum_{j_1=1}^{\left\lfloor \frac{C}{m_1} \right\rfloor} \sum_{j_2=0}^{\left\lfloor \frac{C}{m_2} \right\rfloor} j_1 P(j_1, j_2). \tag{5.24}$$

Using Little's formula, the average time that an s_1 call spends in the system, including waiting as well as service time, is given as $\frac{E(j_1)}{\lambda_1(1 - Pb_1)}$.

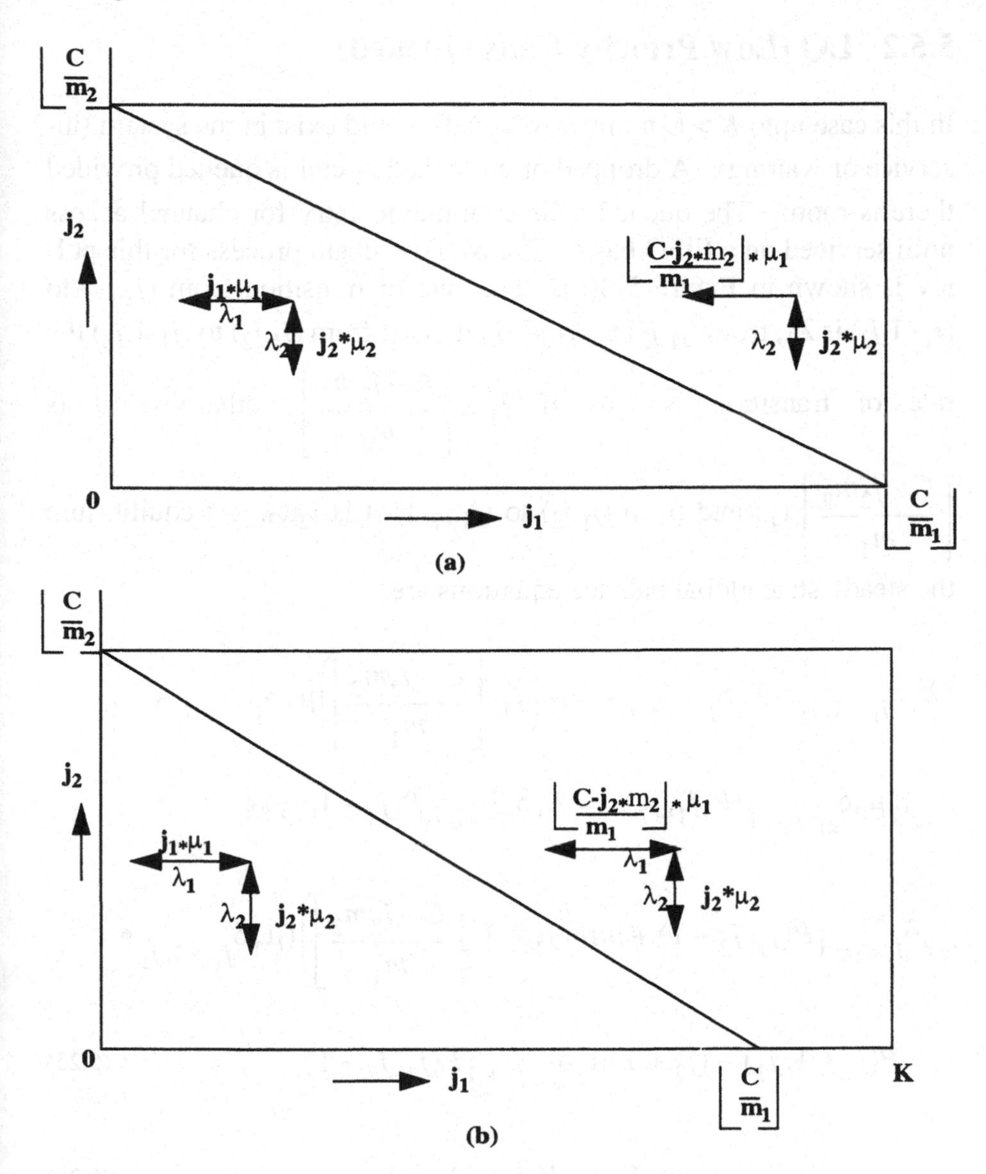

Figure. 5-3. PS with Call Dropping with (a) Dropped Calls Queued (b) Low Priority Calls Queued.

5.5.2 LQ (Low Priority Calls Queued)

In this case upto $K > C$ number of s_1 calls could exist in the system (in-service or waiting). A dropped or a blocked s_1 call is queued provided there is room. The queued calls continue to retry for channel access until serviced on a FIFO basis. The Markov chain process for this policy is shown in Figure 5-3(b). The rate of transition from (j_1,j_2) to (j_1+1,j_2) is λ_1; from (j_1,j_2) to (j_1,j_2+1) it is λ_2; from (j_1,j_2) to (j_1-1,j_2) the rate of transition is $j_1\mu_1$ if $j_1 \le \left\lfloor \frac{C - j_2 m_2}{m_1} \right\rfloor$, otherwise it is $\left\lfloor \frac{C - j_2 m_2}{m_1} \right\rfloor \mu_1$; and from (j_1,j_2) to (j_1,j_2-1) it is $j_2\mu_2$. At equilibrium the steady state global balance equations are:

$$[\lambda_1 \delta_{j_1+1,j_2} + \lambda_2 \delta_{j_1,j_2+1} + min\left(j_1, \left\lfloor \frac{C - j_2 m_2}{m_1} \right\rfloor\right) \mu_1 \delta_{j_1-1,j_2} +$$

$$j_2 \mu_2 \delta_{j_1,j_2-1}] P(j_1,j_2) = \lambda_1 \delta_{j_1-1,j_2} P(j_1-1,j_2) +$$

$${}_2\delta_{j_1,j_2-1} P(j_1,j_2-1) + min\left(j_1+1, \left\lfloor \frac{C - j_2 m_2}{m_1} \right\rfloor\right) (\mu_1 \delta_{j_1+1,j_2} \bullet$$

$$P(j_1+1,j_2) + (j_2+1) \mu_2 \delta_{j_1,j_2+1} P(j_1,j_2+1) \tag{5.25}$$

$$\sum_{(j_1,j_2) \in A} P(j_1,j_2) = 1 \tag{5.26}$$

for all $(j_1,j_2) \in A$. A is the space of acceptable states, which for this case is given as

$$A = \left\{ (j_1, j_2) : (0 \le j_1 \le K) and \left(0 \le j_2 \le \left\lfloor \frac{C}{m_2} \right\rfloor \right) \right\}, \quad (5.27)$$

and δ_{k_1, k_2} is as defined above. Pb_2 is obtained using (5.14), and Pb_1 is given as

$$Pb_1 = \sum_{j_2 = 0}^{\left\lfloor \frac{C}{m_2} \right\rfloor} P(K, j_2). \quad (5.28)$$

The average number of s_1 calls in the system is given as

$$E(j_1) = \sum_{j_1 = 1}^{K} \sum_{j_2 = 0}^{\left\lfloor \frac{C}{m_2} \right\rfloor} j_1 P(j_1, j_2). \quad (5.29)$$

The average system time of an s_1 call is obtained using Little's formula as in the last section.

5.6 DA (PS with Discouraged Arrivals)

In this case the arrival rate of low priority calls is reduced as the number of connections in the link grow. The Markov chain process for this policy is shown in Figure 5-2(b). The rate of transition from (j_1,j_2) to (j_1+1,j_2) is $\left\lfloor \frac{1}{1 + j_1 m_1 + j_2 m_2} \right\rfloor \lambda_1$; from (j_1,j_2) to (j_1,j_2+1) it is λ_2; from (j_1,j_2) to $(j_1\text{-}1,j_2)$ it is $j_1\mu_1$; and from (j_1,j_2) to $(j_1,j_2\text{-}1)$ it is $j_2\mu_2$. At equilibrium the steady state global balance equations are

$$[\lambda_1(j_1, j_2)\delta_{j_1 + 1, j_2} + \lambda_2\delta_{j_1, j_2 + 1} + j_1\mu_1\delta_{j_1 - 1, j_2} +$$
$$j_2\mu_2\delta_{j_1, j_2 - 1}]P(j_1, j_2) = \lambda_1(j_1 - 1, j_2)\delta_{j_1 - 1, j_2}P(j_1 - 1, j_2) +$$

$$\lambda_2 \delta_{j_1, j_2 - 1} P(j_1, j_2 - 1) + (j_1 + 1)\mu_1 \delta_{j_1 + 1, j_2} P(j_1 + 1, j_2) +$$

$$(j_2 + 1)\mu_2 \delta_{j_1, j_2 + 1} P(j_1, j_2 + 1) \tag{5.30}$$

$$\sum_{(j_1, j_2) \in A} P(j_1, j_2) = 1 \tag{5.31}$$

for all $(j_1, j_2) \in A$, where A is as defined in (5.3) and $\lambda_1(k_1, k_2) = \left\lfloor \frac{1}{1 + k_1 m_1 + k_2 m_2} \right\rfloor \lambda_1$. Again, the state probabilities $P(j_1, j_2)$ are obtained by solving (5.30) and (5.31), and the blocking probabilities are obtained by substituting $P(j_1, j_2)$ in (5.6) and (5.7).

The discouraged arrivals scheme described above could be implemented in at least two ways. One way is to accept a call with a probability $\left\lfloor \frac{1}{1 + j_1 m_1 + j_2 m_2} \right\rfloor$ with (j_1, j_2) being the link state at that time. It is similar to the PRP (Probabilistic Reservation Policy) call admission control policy analyzed by Qian *et al.* [36], except that the probability of acceptance of a call is state dependent. The other way is to implement this policy as a part of alternate path routing *i.e.*, the switch overflows s_1 calls to an alternate route at a rate $\left(1 - \left\lfloor \frac{1}{1 + j_1 m_1 + j_2 m_2} \right\rfloor\right) \lambda_1$ where (j_1, j_2) is the state of the primary route.

5.7 Numerical Results

A comparative evaluation of the bandwidth access control policies, discussed above, is performed, in this section, using numerical examples. A single link with two competing services, *i.e.*, $I = 2$, is simulated. The overall capacity of the link is assumed to be 48 units of bandwidth. The other parameters such as (m_1,m_2), (λ_1,λ_2), and (μ_1,μ_2) are taken as (1,6), (120/121,1/121), and (1,0.05) respectively. Similar values for these parameters were chosen in [36]. The mean call holding time of s_2 calls is 20 times larger than that of s_1 calls, whereas, the bandwidth requirement of an s_2 call is 6 times larger than that of an s_1 call. The call arrival rates (λ_1,λ_2) of both services are proportionally increased, and the effect of this increase in traffic on the call blocking probabilities is observed. Each source thus has equally weighted traffic load of

$$\frac{m_i\lambda_i}{\mu_i}.$$

Firstly, the CS policy is assumed to be implemented in the link. The results are presented in Figure 5-4. As expected, due to high ratio of m_2/m_1 CS exhibits unfairness. The narrow-band service s_1 monopolizes the available bandwidth, while starving the wideband service s_2. This effect becomes even more profound as the call arrival rates of both services increase. The CP policy is implemented next, where (C_1,C_2) are arbitrarily taken as (12,36) respectively. The blocking probability of s_2, in this case, improves significantly. The total link throughput, however, will deteriorate under unbalanced traffic load conditions. Figure 5-4 also depicts the PS case with (C_1,C_2) being (18,12) respectively, and C_s being 18 channels. It is apparent from the results of Figure 5-4 that the PS policy can be used to optimize the call blocking probability and throughput characteristics, by fine tuning the shared and engineered bandwidth. Determination of accurate sizes of shared and engineering capacities is anything but trivial as the networks and services are always evolving. An approach to determine the optimum sizes of shared and engineered capacities, iteratively, has been outlined in the next chapter. The other alternative is to use the priority based schemes. The priority based schemes achieve better gains in throughput for higher priority services. This is evident from Figure 5-4 where the blocking probability of higher priority service s_2 is lower than the ones achieved through the conventional CS, CP or PS policies. The blocking probability of s_1, for PS with Discouraged Arrival case, in Figure 5-4 (a), is not included because the discouraged calls do not necessarily mean blocked calls, as explained in Section 5.6.

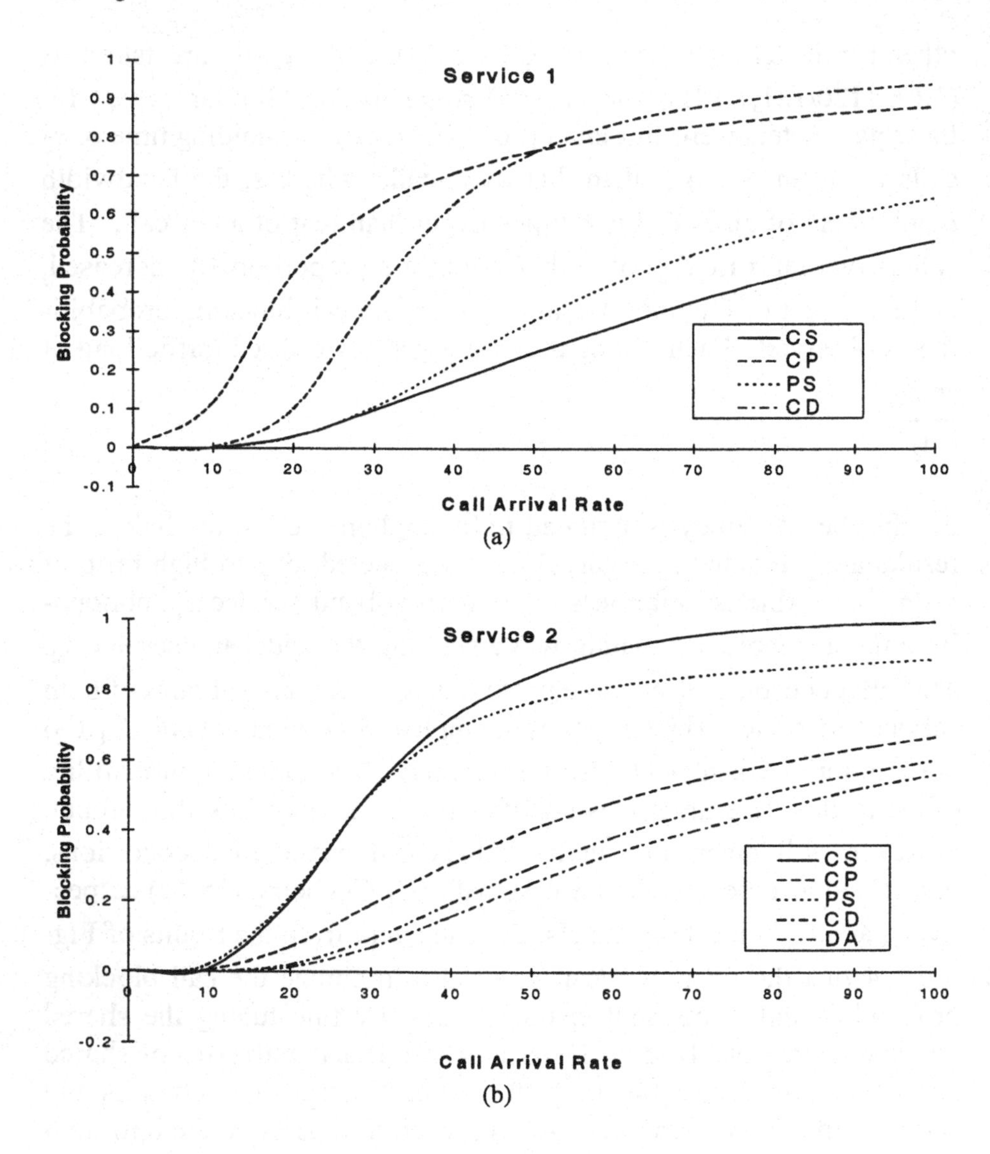

Figure. 5-4. (a) Blocking Probability of Service 1 for CS (Complete Sharing), CP (Complete Partitioning), PS (Partial Sharing), and CD (Call Dropping). (b) Blocking Probability of Service 2 for CS, CP, PS, CD, and DA (Discouraged Arrivals).

Figure 5-5 illustrates the impact of CD, DQ and LQ policies on the call blocking probability of s_1. For LQ, the value of K is chosen to be 60 *i.e.*, at most 60 s_1 calls can stay in the system, either in service or in waiting. As compared to CD, the call blocking probability of s_1 under DQ is higher. The rationale behind this behaviour is that under DQ the dropped calls continue to stay in the system and block new s_1 arrivals that increases the call blocking probability. Also, as expected, LQ results in lower call blocking probability for s_1 at the expense of extra waiting time in the queue. The reason is that under LQ a higher fraction of total incoming calls is accepted because of the queueing, thus reducing the call blocking probability. Since s_2 has a preemptive priority over s_1, its blocking probability will not be effected even if the CD policy is replaced with either DQ or LQ.

Figure 5-6(a) presents the call dropping probability of s_1 calls for the corresponding call arrival rates. The average time spent by s_1 calls, in the system, under DQ and LQ policies is shown in Figure 5-6(b) and Figure 5-6(c), respectively. The peculiar behaviour of call dropping probability of s_1 in Figure 5-6 (a) could be explained as follows. If the call arrival rate of s_2 is very low, then the probability that an s_1 call is dropped in order to make room for an incoming s_2 call is also low. Thus, as the call arrival rate of s_2 increases, the dropping probability of s_1 also increases. On the other extreme, if the call arrival rate of s_2 is very high then the probability that an s_1 call gets accepted, and subsequently dropped, in a link mostly occupied by long lasting s_2 calls is also low. Thus, as illustrated in Figure 5-6(a), the dropping probability of s_1 first increases and then decreases if we continue to increase the call arrival rate of s_2. Similar behaviour can also be observed in Figure 5-7 and Figure 5-8(c) where a different set of traffic and system parameters are chosen.

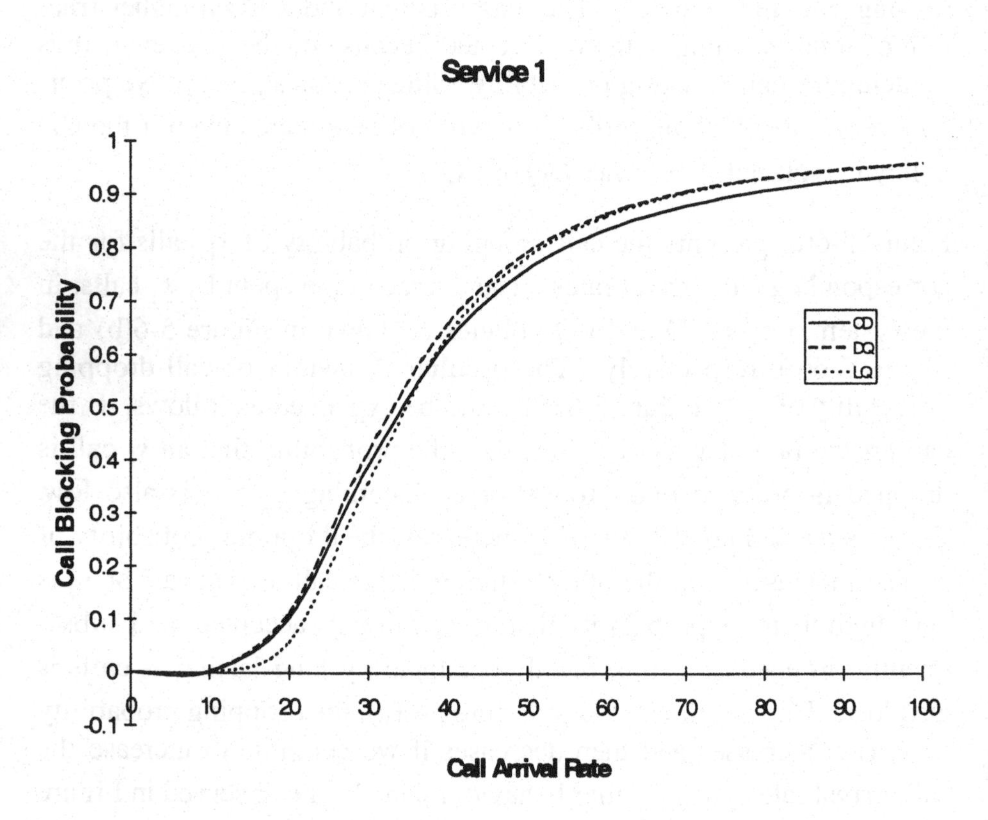

Figure. 5-5. Call Blocking Probability of s_1 under CD, DQ and LQ policies.

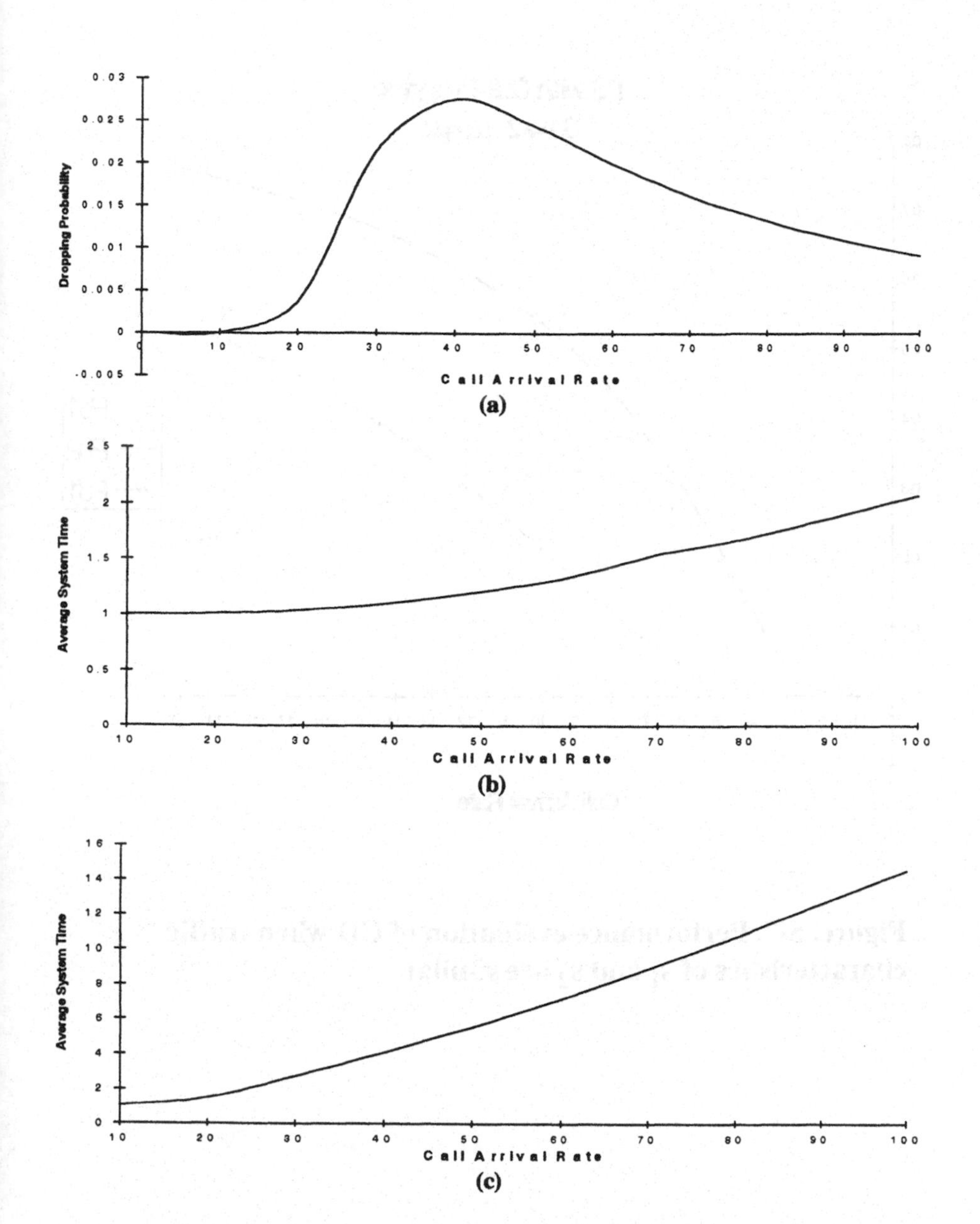

Figure. 5-6. (a) Call Dropping Probability of s_1 calls under PS with Call Dropping scheme (b) Average System Time of dropped calls under Dropped-Calls-Queued scheme (c) Average System Time of low priority calls under Low-Priority-Calls-Queued scheme with K=60.

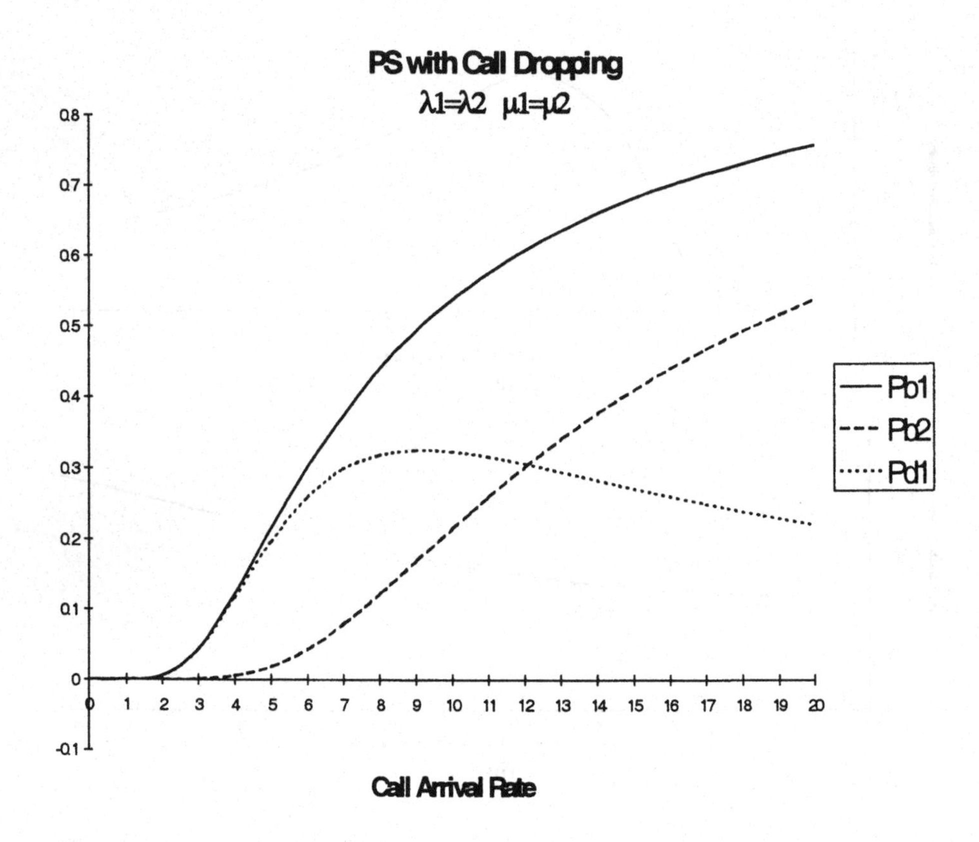

Figure. 5-7. Performance evaluation of CD when traffic characteristics of s_1 and s_2 are similar.

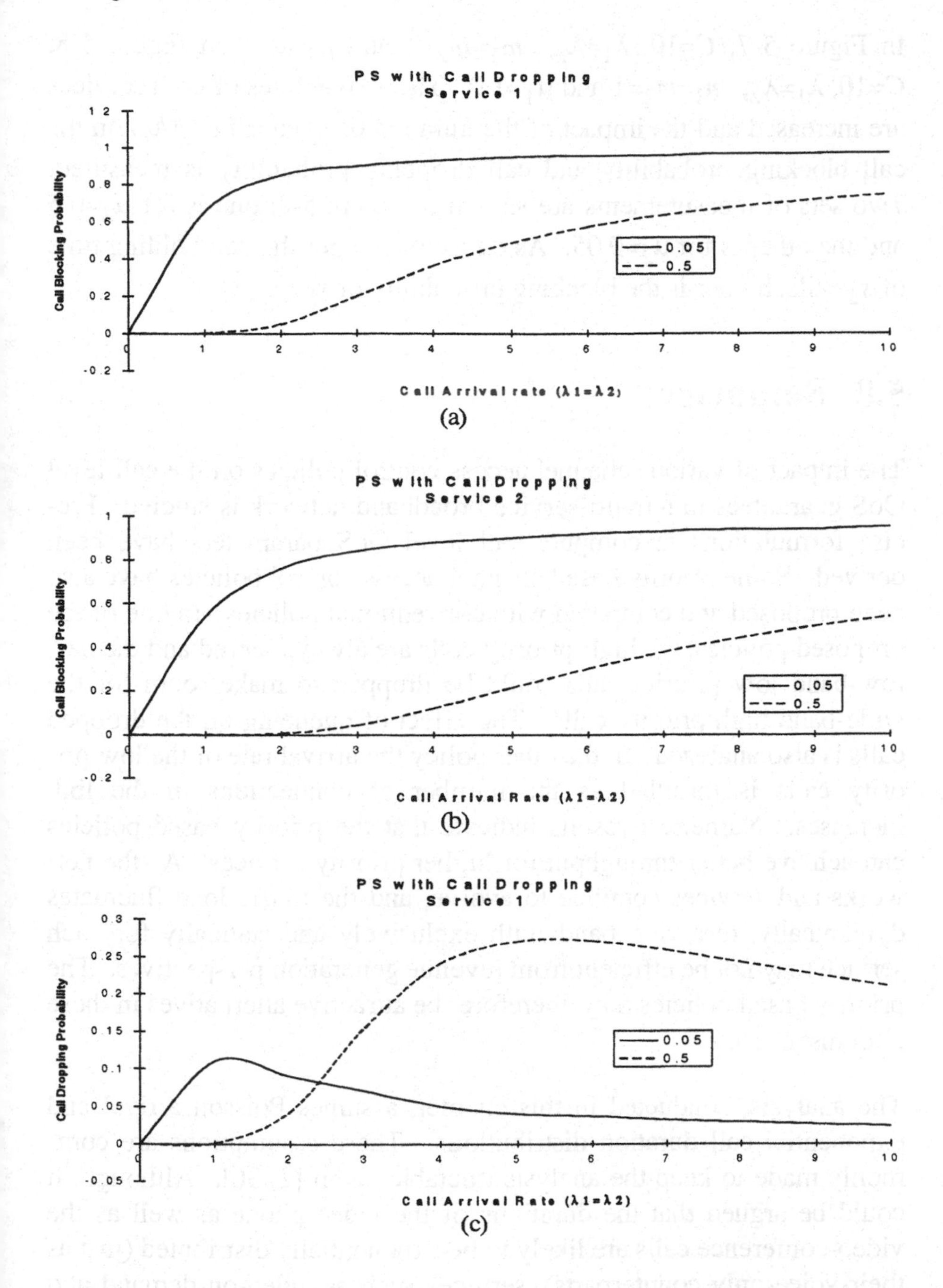

Figure. 5-8. Performance evaluation of CD (a) Call Blocking Probability of s_1 (b) Call Blocking Probability of s_2 (c) Call Dropping Probability of s_1.

In Figure 5-7, C=10, $\lambda_1=\lambda_2$, $m_1=m_2=1$ and $\mu_1=\mu_2$. In Figure 5-8, C=10, $\lambda_1=\lambda_2$, $m_1=m_2=1$ and $\mu_1=1.0$. The arrival rates of both services are increased and the impact of the duration of s_2 calls *i.e.*, $1/\mu_2$ on the call blocking probability and call dropping probability is measured. Two sets of measurements are shown in Figure 5-8; one is for $\mu_2=0.5$ and the other is for $\mu_2=0.05$. As expected, longer the call holding time of s_2 calls, higher is the blocking probability of s_1.

5.8 Summary

The impact of various channel access control policies on the call level QoS guarantees in a multi-service broadband network is studied. Precise formulations to compute call level QoS parameters have been derived. Some priority based channel access control policies have also been proposed and compared with conventional policies. In one of the proposed policies, the high-priority calls are always served and the narrow-band low priority calls could be dropped to make room for the wide-band high priority calls. The effect of queueing on the dropped calls is also analyzed. In the other policy the arrival rate of the low priority calls is throttled as the number of connections in the link increases. Numerical results indicate that the priority based policies can achieve better throughput for higher priority services. As the networks and services continue to evolve, and the traffic load fluctuates dynamically, reserving bandwidth exclusively and statically for each service may not be efficient from revenue generation perspectives. The priority based policies may, therefore, be attractive alternatives in these circumstances.

The analysis, conducted in this chapter, assumes Poisson arrival and exponential call duration distributions. These assumptions are commonly made to keep the analysis tractable, as in [2, 36]. Although, it could be argued that the durations of the video-phone as well as the video-conference calls are likely to be exponentially distributed (just as their voice-only counterparts), services such as video-on-demand and web-browsing may not be exponential. The systems with non-exponential service times, however, can still be analyzed under the framework of a multi-dimensional continuous-time Markov process using approximation techniques (Appendix B). It should, however, be noted that the product form solution, discussed in Section 5.2, is independent of the call duration distribution [18]. Thus, for CS, CP and PS policies, the analysis conducted in this chapter is valid even if the call duration has a non-exponential distribution. The Erlang B formula, used in Section 5.5 for computing the blocking probability of the high-priority service, is also valid for non-exponential call duration distributions [5]. The performance evaluation of multi-service delay-delay and mixed loss-delay systems with non-exponential call holding time distributions is not tractable, and is usually conducted through simulations or using approximations [57].

Based on the analysis of this chapter, a bandwidth optimization framework for multi-service broadband networks is developed next. In the aforesaid framework, depending upon the anticipated traffic load, a suitable channel access control policy is determined for each link in the network, such that the overall network revenue is maximized while ensuring the desired QoS.

Chapter 6

Bandwidth Optimization in Broadband Networks

In this chapter, firstly, the QoS in the entire network is analytically determined; and secondly, a heuristics based bandwidth optimization procedure is outlined. Given a set of SD (Source-Destination) pairs, and a list of routes for each of the SD pairs, the procedure develops a channel allocation plan that optimizes QoS and revenue.

As mentioned in Chapter 1, network optimization has been the subject of numerous studies including [3], [20], [23], and [33], under a variety of contexts. Notable among these is the approach by Kheradpir *et al.* [23] where routing and bandwidth allocation has been jointly addressed for a SONET network using an integer programming based technique. The heuristics based approach, proposed in this chapter, however, has at least two advantages over such integer programming based techniques. One is that the proposed approach is incremental. For minor variations in traffic, involving only a few SD pairs, the proposed procedure will achieve an optimal solution within a few iterations by conducting a search for the optimal solution in the vicinity of the affected paths. The second advantage is the flexibility. The proposed heuristics based framework is quite flexible in the sense that it can accommodate networks with a heterogeneous mix of channel access control policies, implemented non-uniformly at the switches in the network. The integer programming based techniques are effective only for the networks that support simple policies such as CS, CP and PS. It is quite difficult to specify the complex channel access control criteria such as CD and DA to an integer programming framework.

6.1 The Network Model

Consider the model of a network as shown in Figure 6-1. At each source node, multiple paths are available to each SD pair. Each path is an ordered list of transmission links. The transmission links are assumed to be independent. An SD pair may be associated with one or multiple services. λ_i^r is the aggregate arrival rate from the SD pair r belonging to service s_i. A call may be accepted or denied based on the channel access control policy of each link.

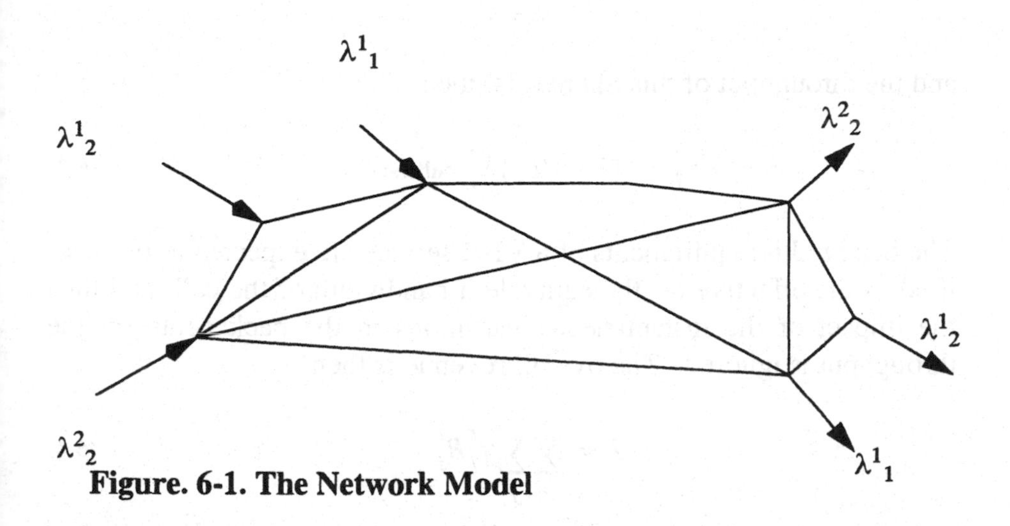

Figure. 6-1. The Network Model

At first the blocking probabilities are computed at each link in the network based on the anticipated traffic load. Thereafter, the blocking probabilities of each SD pair are computed. The results of these QoS tests are, thereafter, used to trigger and drive the bandwidth optimization procedure. The traffic descriptors such as the average offered call rate, average carried call rate and average call holding time for each SD pair of each service are estimated by periodically reading the operational measurements from the switches. The optimization procedure is invoked if the offered average call arrival rate is outside the specified thresholds.

The call blocking probability of an SD pair r associated with service s_i along a route k is given by

$$Pb_i^{rk} = 1 - \prod_l (1 - Pb_i^l) \tag{6.1}$$

where l are the links along this route, and Pb_i^l is the blocking probability of service s_i at the link l, as derived in the last chapter. Given that there are K disjoint routes available to this SD pair, the overall blocking probability of this SD pair along all its K paths is then given by

$$Pb_i^r = \prod_{k=1}^{K} Pb_i^{rk} \tag{6.2}$$

and the throughput of this SD pair is, then

$$\gamma_i^r = (1 - Pb_i^r)\lambda_i^r \text{ calls/sec.} \tag{6.3}$$

The bandwidth requirements of a VBR service are expected to be specified as the effective or the equivalent bandwidth of the call, and thus the impact of the instantaneous variations in the packet-rate on the throughput is ignored. The overall revenue is then

$$J = \sum_i \sum_r \gamma_i^r R_i^r \tag{6.4}$$

where R_i is the average revenue generated when an s_i type call from a particular SD pair gets accepted. The cost of resource utilization due to accepted calls is not considered to keep the revenue function simple.

6.2 Bandwidth Optimization Procedure

A heuristics based approach is used to find a global solution for bandwidth allocation. Upon detecting potential violations of QoS guarantees, as a result of anticipated fluctuations in traffic, the algorithm goes through a series of iterations; and at each iteration, it attempts to alleviate or minimize the problem. Firstly, the algorithm considers accommodating the excess load by altering the channel access control policies in such a way that the QoS of the affected SD pairs improves while ensuring that the QoS guarantees of others are not tempered with. Secondly, more bandwidth, if available, is allocated to the affected SD pairs. Lastly, the algorithm further fine tunes the channel access control policies to maximize the revenue.

The redistribution of resources is done in small BBUs (Basic Bandwidth Units). At each step, blocking probabilities of the affected SD pairs as well as the network revenue are computed, using (6.2) and (6.4) respectively. Only those PCs (Potential Configurations) are considered for actual implementation that constrain the blocking probabilities within bounds and maximize the network revenue. The algorithm is outlined below. For simplicity only CS, CP and PS policies are considered, and the QoS is specified in terms of call blocking probability. The procedure could be easily expanded to accommodate other channel access control policies such as CD and DA, discussed in the last chapter.

Input:

- a set of disjoint routes available to each SD pair
- traffic estimates for each SD pair
- QoS guarantees *i.e.*, maximum acceptable call blocking probability for each SD pair
- network topology
- existing channel access control policy for each link

Output: A channel allocation plan for the entire network that optimizes QoS and revenue.

Step 1: Check if the predicted increase in load, or a portion of it, can simply be accommodated through proper load balancing.

For each affected SD pair *Do* *{loop 1}*

For each specified route of the SD pair *Do* *{loop 2}*

For each specified link in the route *Do* *{loop 3}*

consider borrowing 1BBU as shared or engineered bandwidth

compute the blocking probabilities of SD pairs that are influenced by this change

If the overall blocking probability of the SD pairs reduce then save this as a PC (Potential Configuration).

From the PC set, keep the PC that reduces blocking probability the most, and discard others. Consider this PC to be permanent.

If the blocking probabilities of all SD pairs is below the maximum acceptable then go to Step 3. If the PC set is empty and there are still some SD pairs with QoS violations then go to Step 2 as more bandwidth is needed. Otherwise continue repeating loop 1.

Step 2: Check if the predicted increase in load, or a portion of it, can be accommodated by allocating available free bandwidth

Start from the network configuration resulting from Iteration 1

Consider only those affected SD pairs from Iteration 1 whose QoS is still in violation

For each affected SD pair *Do* *{loop 1}*

For each specified route of the SD pair *Do* *{loop 2}*

For each specified link of the route *Do* *{loop 3}*

consider allocating 1 BBU as shared or engineered capacity

compute the blocking probabilities of SD pairs that are influenced by this change

save this as a PC.

From the PC set, select the PC that reduces the blocking probability of the SD pairs the most.

Continue repeating loop 1 until blocking probabilities of all SD pairs are within the constraints. If no free bandwidth is available then EXIT.

Step 3: Maximize Revenue.

For each SD pair *Do* *{loop 1}*

For each specified route of the SD pair *Do* *{loop 2}*

For each specified link in the route *Do* *{loop 3}*

consider borrowing 1BBU as shared or engineered bandwidth

save this as a PC if there is no violation of QoS

From the PC set, select the PC that improves revenue the most and discard others.

If none found then EXIT as no more gain in revenue is feasible; else continue repeating loop 1.

Example:

Two objective functions are involved in the above optimization. One is the network revenue computed using (6.4), that is to be maximized. The other is the blocking probability of each SD pair that should be less than the maximum acceptable blocking probability of the associated services *i.e.*, $Pb_i^r \leq MaxPb_i$. Consider as an example a single link shared by two SD pairs, each belonging to a different service. Let (m_1,m_2), (λ_1,λ_2), (μ_1,μ_2), $(MaxPb_1, MaxPb_2)$, and (R_1,R_2) be (1,6), (20*120/121,20*1/121), (1,0.05), (0.2,0.2) and (120,1) respectively. The superscripts are ignored for clarity. The total number of channels available on this link *i.e.*, C = 48 channels, and 1BBU = 6 channels. Let the initial configuration *i.e.*, (C_1,C_2,C_s) be (24,24,0), where C_1 and C_2 are the engineered bandwidths for SD pair 1 and 2 respectively and C_s is the shared bandwidth. These values render (Pb_1, Pb_2) as

(0.06321, 0.23942) with network revenue $J = 33.65847$. Clearly, there is QoS violation for s_2. The algorithm then works as follows:

Step 1

Iteration 1: The PCs are (18,24,6), (18,30,0), (24,18,6), and (30,18,0). The PC (18,24,6) is selected that produces (Pb_1, Pb_2) as (0.073718, 0.187939) with $J = 34.47$. Since $Pb_1 < MaxPb_1$ and $Pb_2 < MaxPb_2$, the objective of this step is accomplished and there is no need for Step 2. The algorithm goes straight to Step 3 to see if the revenue could be maximized without violating QoS.

Step 3

Iteration 1: The PC (12,24,12) is selected that produces (Pb_1, Pb_2) as (0.075210, 0.183937) with $J = 34.5204 > 34.47$ (from last iteration).

Iteration 2: The PC (6,24,18) is selected that produces (Pb_1, Pb_2) as (0.0752186, 0.1839208) with $J = 34.52062 > 34.5204$ (from last iteration).

Iteration 3: No PC could produce $J > 34.52062$ (from last iteration) indicating that revenue can not be improved any further.

For any initial configuration, the algorithm always converges to (6,24,18), provided QoS constraints are the same as above. This is because the revenue function for the above two service types and given traffic loads has a clear global maxima, as shown in Figure 6-2. As the weighted traffic loads are equal *i.e.*, $\frac{m_1 \lambda_1}{\mu_1} = \frac{m_2 \lambda_2}{\mu_2}$, the maxima is achieved when $C_1 = C_2$ and $C_s = C$. The maxima exists because of certain properties of Pb_i, which are discussed below.

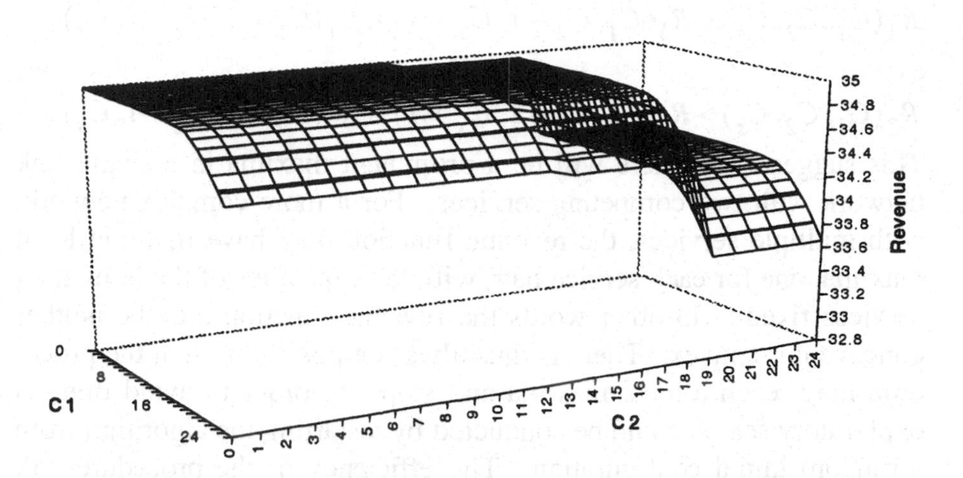

Figure. 6-2. Network revenue as a function of (C_1, C_2, C_s) for a two service single link case with specified traffic load.

Given the traffic parameters m_i, λ_i, and μ_i, the blocking probability of a service is a function of the available channels. We will, therefore, represent Pb_i as $Pb_i(C_1, C_2, ..., C_I, C_s)$, for now. For a single link, with only one service, the blocking probability of the service is obtained by using the Erlang-B formula, which is a convex function and strictly decreasing in the number of channels [5, 54]. For a single link, with two competing services, it can be easily proved that $Pb_1(C_1, C_2, C_s) > Pb_1(C_1, C_2, C_s + 1) > Pb_1(C_1 + 1, C_2, C_s)$ *i.e.*, allocating an additional BBU improves the blocking probability of a service. The improvement is higher if it is allocated exclusively to the service rather than as a shared capacity. Consequently, $R_1(C_1, C_2, C_s) < R_1(C_1, C_2, C_s + 1) < R_1(C_1, C_2 - 1, C_s)$, where R is the average revenue generated by the service in a single link network. Also, if $C_1 + C_2 + C_s = C$ then $Pb_1(C_1, C_2, C_s) > Pb_1(C_1, C_2 - 1, C_s + 1) > Pb_1(C_1 + 1, C_2 - 1, C_s)$, as well as, $Pb_2(C_1, C_2, C_s) < Pb_2(C_1, C_2 - 1, C_s + 1) < Pb_2(C_1 + 1, C_2 - 1, C_s)$ Correspondingly,

$R_1(C_1, C_2, C_s) < R_1(C_1, C_2 - 1, C_s + 1) < R_1(C_1 + 1, C_2 - 1, C_s)$, as well as, $R_2(C_1, C_2, C_s) > R_2(C_1, C_2 - 1, C_s + 1) > R_2(C_1 + 1, C_2 - 1, C_s)$. This suggests that there will be a prominent maxima in a single link network with two competing services. For a more complex network, with multiple services, the revenue function may have multiple local maxima, one for each service pair, with the capacities of the remaining services fixed. In other words the revenue function may be neither concave nor convex. There is thus always a possibility that the procedure may reach a local maxima and stop. In order to avoid this, an exploratory search could be conducted by restarting the algorithm from a random initial configuration. The efficiency of the procedure will also depend on the size of the BBU. A smaller BBU, on one hand, will improve the accuracy of the bandwidth allocations but, on the other hand, will also increase the computational complexity of the procedure as well as the possibility of discovering a local maxima instead of a global.

6.3 Summary

The problem of adaptively managing the bandwidth of a multi-service broadband network by implementing optimal channel access control policies at each link is studied. The main contribution of this chapter is the formulation of this problem as an optimization of network revenue and QoS among competing services. The procedure proposed for the said optimization is simple and intuitive.

The algorithm needs to be tested on a real network using actual traffic traces. Besides, the proposed framework needs to be thoroughly compared to other approaches such as integer programming based techniques. The effectiveness of the algorithm in achieving a maxima, however, could be verified intuitively, as done in Section 6.2 using an example. Some other aspects that need further investigation include the assumption of independence among network links, made herein, which generally holds for mesh topologies under heavy loaded conditions, but may not be true under lightly loaded conditions. Secondly, it is assumed here that all the possible routes of an SD pair are known.

The problem of optimal routing is, therefore, not addressed. Lastly, it is apparent that the performance of the algorithm depends considerably on the computational complexities of the channel access control policies. For the coordinate convex policies such as CS, CP and PS, the product form solution is applicable which is computationaly less intensive [5, 18]. For other policies, however, computationaly intensive numerical methods may be needed to compute the QoS parameters that will deteriorate the run-time performance of the algorithm. It should be noted, though, that the strategy proposed in this chapter is not for real-time control but for short to mid-term bandwidth management. It is to be used by a network level controller to periodically estimate the bandwidth requirements, and subsequently pass the estimated parameters to the network switches where the real-time call admission control and/or congestion control procedures are implemented. The call and packet level controllers thereafter function within the bounds of these computed parameters.

In the next chapter the above framework is extended to the cellular networks.

Chapter 7

Capacity Optimization in Cellular Networks

In this chapter, the framework of capacity optimization proposed in the last chapter is extended to multimedia cellular networks; and two main aspects of cellular networks are analyzed. One of the said aspects is the user mobility; while the second aspect is the soft capacity of cells in DS-CDMA based cellular networks, that needs to be taken into account for QoS evaluation and capacity requirements estimation.

Firstly, user mobility is modeled and integrated with the call model. This model is then used to evaluate QoS in cellular networks. Based on the analysis by Markoulidakis *et al.* [32], and Xie and Kuek [53], we conjecture herein that the multi-service analysis of Chapter 5 is directly applicable to cellular networks with the assumptions that the call originations in a cell as well as the handover arrivals are Poisson; and the channel holding time in a cell is exponential. Subsequently, the bandwidth optimization procedure, outlined in Chapter 6 for wireline networks, is adapted for cellular networks. To the best of our knowledge, such generalized framework for capacity optimization in multi-service cellular networks has not been previously presented in literature. The issue of the soft cell capacity in CDMA networks is also addressed. Towards that end, a new analytical model is developed that characterizes resource consumption by VBR (Variable Bit Rate) video sources. Bursty VBR traffic is assumed to be handled using multiple parallel channels as proposed in recent CDMA standards [24, 34]. This model, coupled with a similar voice model, is used to analytically estimate statistical multiplexing gain and soft capacity in terms of bit error rate in a DS-CDMA cell. The performance of various channel access control policies such as CS (Complete Sharing), CP (Complete Partitioning) and PS (Partial Sharing) is re-evaluated hereunder by taking into consideration the said aspects of cellular networks. As mentioned in Chapter 1, although the capacity estimation and allocation aspects have been addressed in [10], [11], [14], [15], [28], [46], and [55], most of these approaches ignore user mobility and VBR traffic.

7.1 Network Model

A two dimensional cellular grid with ideal propagation is assumed. The propagation modeling is outside the scope of this work. The average call origination rate is considered to be independent of the number of calls in progress, because of the high subscribership. The network may have stationary as well as mobile users. A mobile is served by the base station in its current cell. When the mobile crosses a cell boundary into an adjacent cell while the call is in progress, a handover procedure takes place. In FDMA/TDMA based cellular networks, if no channel is available in the new cell into which the mobile moves, the handover call is forced to terminate before completion. In CDMA based cellular networks, there are no hard channels due to which a handover call is dropped before completion if the admission of the call causes the bit error rate in the new cell to deteriorate below some threshold.

Let C be the total channels available in a cell n, and that each call belonging to s_i occupies m_i channels for the duration of the call. For VBR sources m_i is either the peak rate or the equivalent bandwidth of the call. In FDMA/TDMA the capacity of the cell is fixed while in CDMA the cell capacity C depends on the SINR (Signal to Noise and Interference Ratio) requirements of the sources. The impact of soft capacity in CDMA is discussed later in this chapter. The traffic corresponding to service s_i is assumed to arrive in a cell at a Poisson rate λ_i with exponentially distributed channel holding time of mean $\frac{1}{\mu_i}$. It may be noted that $\lambda_i = \lambda_{i_o} + \lambda_{i_h}$, where λ_{i_o} is the rate at which new calls belonging to service s_i originate in the cell; and λ_{i_h} is the handover call arrival rate for that service in that particular cell. Both are assumed to be Poisson. The average channel service rate μ_i is the rate at which the carried calls belonging to service s_i are completed in the cell or are handed off to the other cells. Thus, $\mu_i = \mu_{i_d} + \mu_h$, where $1/\mu_{i_d}$ is the mean call holding time given that either the call terminates in the cell it originated or no handover fails, and μ_h is the outgoing handover rate per terminal [32, 53]. The state of the cell can then be modeled as an I-dimensional Birth-Death Markov process, with vector $\boldsymbol{j} = \{j_1, j_2, \ldots, j_i, \ldots, j_I\}$ representing the number of connections from each service that are active in the cell at a given time. The objective now is to determine the probability of being in state $\boldsymbol{j}$ *i.e.*, $P(\boldsymbol{j})$, and then compute the call blocking probability and the call dropping probability for each service, s_i, first at the cell level and, thereafter, at the network level. The blocking probability Pb_i is the probability that a call belonging to service s_i will be blocked in the network. A BCC (Blocked Calls Cleared) system is assumed. The call dropping probability Pd_i is the probability that a non-blocked call is dropped due to handover failure. The QoS of s_i is then better characterized as $QoS = \alpha Pb_i + (1-\alpha)Pd_i$ where $\alpha < 1$. Depending upon the relative importance of call blocking probability and call dropping probability for a service, a particular channel allocation plan could be deduced, or an existing one could be fine tuned. The cell state $\boldsymbol{j}$ depends on the channel allocation policy implemented in the cell. Assuming that the call arrival process is stationary, $P(\boldsymbol{j})$ is obtained by solving the steady-state equilibrium equations of the I-dimensional Birth-Death Markov process. Given a particular channel allocation policy and the traffic intensity $\rho_i = \frac{\lambda_i}{\mu_i}$, the blocking probability Pb_i of service s_i in cell n is derived as in Chapter 5.

It could be argued that in the networks with small size cells the exponential channel holding time distribution may not hold. The computational benefit of product form solution, for even the coordinate convex policies such as CS, CP and PS, thus may not exist. It is however pointed out in [5], and [18] that the product form solution is valid for a wide spectrum of holding time distributions, extending from the class of distributions possessing rational Laplace transform to general distribution. Thus, in small cell networks, the product form solution, used in Chapter 5 for CS, CP and PS policies, would still be valid provided the channel holding time distribution in these networks is among the aforementioned distributions.

Once the QoS per cell is known, the QoS of the entire service area or network is evaluated. Assuming that the events in the cells are mutually independent, the probability that a call belonging to a service s_i is blocked in a service area with a total of N cells is obtained as

$$Pb_i = \frac{\sum_{n=1}^{N} \lambda_i^n \cdot Pb_i^n}{\sum_{n=1}^{N} \lambda_i^n}. \tag{7.1}$$

Pb_i^n is the blocking probability of service s_i in a cell n based on the particular channel allocation policy implemented at that cell site. The overall throughput of a service is then

$$\gamma_i = \sum_{n=1}^{N} \lambda_i^n \cdot (1 - Pb_i^n). \tag{7.2}$$

The net average revenue can be expressed as

$$J = \sum_{i=1}^{I} \gamma_i \cdot R_i \tag{7.3}$$

where R_i is the average revenue generated per s_i call.

The call dropping probability is determined as follows:

The memoryless property of the exponential distribution of channel holding time and call holding time implies that the duration of a call after the handover has same distribution as before the handover. The probability that handover occurs before the call completes is, therefore, given as (Appendix B):

$$h_i^n = \frac{\mu_h^n}{\mu_{i_d} + \mu_h^n} \tag{7.4}$$

where μ_{i_d} is the mean call duration of s_i calls, and μ_h^n is the average handover rate in cell n [49]. Analytical models have been proposed by Hong and Rappaport [19], and Thomas *et al.* [49] to compute the average handover rate based on user mobility and cell geometry. It is assumed here that these values are determined empirically through network traffic monitoring. Let $P_i^{n_1 n_2}$ be the probability that a call from cell n_1 is handed off to the neighboring cell n_2, also determined empirically through network traffic monitoring. Assuming that a call can originate from any cell in the service area, the probability that a call belonging to service s_i is not complete and is dropped before the kth handover, where k=1,2,3..., is given by

$$Ph_i(k) = \frac{1}{N} \sum_{\{n_0, n_1, \ldots, n_k\}} h_i^{n_0} P_i^{n_0 n_1} \left(1 - Pb_i^{n_1}\right) \cdot h_i^{n_1} P_i^{n_1 n_2} \left(1 - Pb_i^{n_2}\right) \ldots$$

$$h_i^{n_{k-1}} P_i^{n_{k-1} n_k} Pb_i^{n_k}$$ for all $\{n_0, n_1, \ldots n_k\}$ such that (n_i, n_{i+1}) are neighboring cells. (7.5)

The sequence $\{n_0, n_1, \ldots n_k\}$ is the sequence of cells that a mobile may traverse during its k handovers, with round trips allowed. The dropping probability is then

$$Pd_i = \sum_{k=1}^{\infty} Ph_i(k). \tag{7.6}$$

However if we assume a uniform traffic distribution with $h_i^n = h_i$ and equal transition probabilities then (7.5) simplifies to

$$Ph_i(k) = [h_i(1 - Pb_i)]^{k-1} h_i Pb_i \tag{7.7}$$

and the dropping probability is then

$$Pd_i = \sum_{k=1}^{\infty} Ph_i(k) = \sum_{k=1}^{\infty} [h_i(1 - Pb_i)]^{k-1} h_i Pb_i = \frac{h_i Pb_i}{1 - [h_i(1 - Pb_i)]} \tag{7.8}$$

7.2 Capacity Optimization Procedure

A heuristics based approach, similar to the one proposed in chapter 6, is used to find a global solution for capacity dimensioning in wireless cellular networks. Upon detecting potential violations of QoS guarantees for particular services in particular cells, as a result of anticipated fluctuations in traffic, the algorithm goes through a series of iterations, and at each iteration, it attempts to alleviate or minimize the problem. Firstly, the algorithm considers redistributing channel allocations within the cell, while ensuring that QoS guarantees of other services are not tampered with. Secondly, channels from neighboring cells are borrowed, as long as QoS is not violated in those cells. It is apparent that if the subscriber distribution is non-uniform then a uniform channel assignment is not optimal. Thus, the frequency plan should be frequently updated based on the fluctuations in the underlying traffic. Lastly, the algorithm makes attempts to improve the network revenue. The algorithm is as follows:

Input:

- estimates of expected traffic intensities for each service in each cell (call origination as well as handover arrival)
- desired QoS for each service in terms of call blocking and call dropping probabilities. In other words the maximum acceptable call blocking probability and call dropping probability for each service *i.e.*, $MaxPb_i$ and $MaxPd_i$, respectively.
- existing cell layout, frequency and channel allocation plans

Output: An optimal channel allocation plan indicating the number of channels to be reserved for each offered service in each cell.

Step 1: Check if the predicted increase in load, or a portion of it, can simply be accommodated by fine tuning the channel allocation policy of each affected cell. The objective in this step is to minimize $\sum_i max(0, Pb_i - MaxPb_i)$ as well as $\sum_i max(0, Pd_i - MaxPd_i)$ to zero.

For each affected cell *Do* *{loop 1}*

For each affected service *Do* *{loop 2}*

consider borrowing 1 BBU as shared or engineered capacity

compute the QoS of all affected services

If the overall QoS improves then save this as a PC (Potential Configuration).

From the PC set keep the PC that improves QoS the most, and discard others. Consider this PC to be permanent.

If the PC set is empty and QoS is still in violation then go to Step 2 as more capacity is needed. If the QoS is satisfied then go to Step 3. Otherwise continue repeating loop 1.

Step 2: Check if the predicted increase in load, or a portion of it, can be accommodated by borrowing channels from neighboring cells (This iteration is not applicable to CDMA networks as the same spectrum is used in each cell). The objective in this step is to again minimize $\sum_i max(0, Pb_i - MaxPb_i)$ as well as $\sum_i max(0, Pd_i - MaxPd_i)$ to zero.

Start from the channel allocation plan resulting from Iteration 1

Consider only those cells from Iteration 1 where QoS is still in violaion

For each affected cell *Do* *{loop 1}*

For each affected service *Do* *{loop 2}*

From each neighboring cell *Do* *{loop 3}*

consider borrowing a small number of channels as shared or engineered capacity, provided co-channel interference is not violated (by maintaining proper frequency reuse distance)

compute QoS of all affected services

If the QoS improves without violating the QoS in other cells then save this as a PC (Potential Configuration).

From the PC set, keep the PC that improves QoS the most, and discard others. Consider this PC to be permanent.

If the PC set is empty then EXIT as no more gain is likely in QoS; Else continue repeating loop 1 until QoS of all services are within the constraints.

Step 3: The objective in this step is to maximize revenue as defined by (7.3).

For each cell *Do* *{loop 1}*

For each service *Do* *{loop 2}*

consider 1 BBU as shared or engineered capacity

save this as a PC if the revenue improves without violating the QoS.

From the PC set keep the PC that improves revenue the most, and discard others. Consider this PC to be permanent.

If the PC set is empty then EXIT as no more gain in revenue is likely. Otherwise continue repeating loop 1.

The above procedure is further explained using flow charts in Appendix C. It is an adaptation of the bandwidth optimization procedure suggested in Chapter 6. In the cellular networks, the additional constraint is that the cell capacity is not fixed. In TDMA and FDMA based cellular networks, the frequency channels are reused at a regular distance. The distance is kept long enough to guarantee a minimal co-channel interference. This results in a frequency plan that appears as a tiling of cell clusters over the entire service area. The frequency allocations within the clusters may be uniform or non-uniform. If, for example, a cellular network has been granted a band of C^T channels, then each cell in a cluster may have a non-overlapping sub-set of these channels such that $\sum_{m=1}^{M} C_s^m + \left(\sum_{i=1}^{I} C_i^m \right) \le C^T$, where m=1..M, is a cell in a cluster.

On the other hand, in CDMA based cellular networks, even though same frequency band may be reused in every cell, the cell capacity is still not fixed. It is a function of the bit error rate requirements, the number of interfering users within a cell, and the number of interfering users in the neighboring cells. Estimation of CDMA soft cell capacity is further explored in the subsequent sections.

7.3 Soft Capacity of CDMA Cells

Unlike FDMA/TDMA (Frequency/Time Division Multiple Access) where each user is assigned an available frequency/time slot to communicate, the spread-spectrum based DS-CDMA networks assign a unique code to each user. The user information is spread by using the code and transmitted over the shared frequency channel. The same code is used at the receiver's end to de-spread and recover the target signal from the aggregate traffic. As there is no physical limit to the number of codes that could be assigned (provided orthogonality among codes is ensured), the capacity of a CDMA cell also does not have a hard upper limit. The capacity however is interference limited and depends on traffic aggregation and power control. Adding more users to the cell will add to the interference, causing cell outage eventually. The relationship between CDMA cell capacity and SINR (Signal to Interference and Noise Ratio) on the reverse link (subscriber to cell-site), which is considered to be the limiting case in CDMA, is expressed as [54, 9]:

(Number of Users × Signal Power/User) + Other Cell Interference +

Thermal Noise ≤ Total Interference .

With some simplifying assumption [9], the above relation can be expressed as:

$$\frac{E_b}{No} = \frac{W}{B} \cdot \frac{1}{c-1} \cdot \frac{1 - \dfrac{N_{thermal}}{No}}{1+\eta} \qquad (7.9)$$

where $(c - 1)$ is the number of interfering channels, W is the available spectrum for CDMA (for IS-95 it is 1.25 MHz), B is the connection bit rate in bits per second, E_b is the energy in the information bit, No is the noise power, η is the cell overload factor representing the outer-cell interference *i.e.*, the interference from the neighboring cells, and $N_{thermal}$ is the thermal noise. As mentioned earlier, some simplifying assumptions are made while deriving (7.9). For example, in actual practice, the signal power of each user is not constant and varies depending upon the user mobility and power control. A consequence of this is the potential deterioration of SINR during the call. We, however, assume that a perfect power control is implemented in the network which ensures that the received signal power of all the channels in the cell is equal, and remains constant irrespective of the user mobility and location within the cell. Furthermore, the other-cell interference is also random and depends on the users in the neighboring cells. A tight upper bound of 0.55, however, has been determined for η in [9]. At the expense of possible under-utilization, we assume that the other-cell interference η is also fixed at 0.55.

Determination of maximum number of sources that could be aggregated without violating SINR is essential for access control or capacity planning purposes. Analytical models are, therefore, developed herein that characterize the channel utilization by the multimedia sources, which subsequently help quantify the statistical multiplexing gain or extent of traffic aggregation that is feasible in a CDMA cell, with acceptable SINR. Real-time VBR video services are considered, in addition to voice and CBR (Constant Bit Rate) services. Traffic from VBR video sources is modeled as an AR (AutoRegressive) process with Gaussian distribution. Bursty data is assumed to be handled using multiple parallel channels as suggested in IS-95-B and cdma2000 standards [24, 34]. The channel activity parameters are estimated based on the assumptions about the underlying source traffic.

7.3.1 Cell Capacity Estimation

Consider a CDMA cell supporting VBR video, voice as well as CBR sources. Each call is assigned one or multiple channels depending upon the source traffic. For simplicity we assume that the voice and other service types are assigned only one channel whereas the VBR video calls may be assigned an additional channel to handle the bursty traffic. The channels are of equal bit rate, and always transmit at the peak channel rate of B bits/sec. Each of the channels is assigned a code which is orthogonal to any other code, in use, in the cell. A sufficiently large set of orthogonal codes is assumed to be available. It is apparent that CBR (Constant Bit Rate) sources will not contribute to any statistical multiplexing as the channels are continuously active. The statistical multiplexing is, however, possible for the VBR sources because of the channel activity factor. Channel utilization models that characterize the resource consumption by such traffic sources are developed next. These models are subsequently used for estimating the statistical multiplexing gain achievable in the CDMA networks.

7.3.1.1 Channel Utilization Models

a) VBR Video Sources

Consider a digitized and compressed video source involved in a DS-CDMA connection. The codec processes a frame, stores it in internal buffers and then transmits the compressed frame over the next frame interval [12]. Assuming burst mode and a frame rate of F frames/sec, if the size of the compressed frame is less than or equal to $\frac{B}{F}$ bits/frame, then only a single channel is used. If, on the other hand, the size of the compressed frame is greater than $\frac{B}{F}$, then an additional channel is used, in parallel, to transmit the excess bits, as illustrated in Figure 7-1(a). We characterize the utilization of the channels using three parameters *i.e.*, $P^1{}_{on}$, $P^2{}_{on}$, and $P^2{}_{off}$, where $P^1{}_{on}$ is the probability that only one channel is ON; $P^2{}_{on}$ is the probability that both channels are ON; and $P^2{}_{off}$ is the probability that both channels are OFF. In order to determine these parameters we need to know the source traffic characteristics.

As mentioned in Chapter 6, for video-phones and video-conferencing, with only few scene changes and a low activity within the scene, a lower order AR process of (4.1) closely matches the inter-frame bit rate characteristics of these sources. The parameters $P^1{}_{on}$, $P^2{}_{on}$, and $P^2{}_{off}$ could then be estimated as:

$$P^1{}_{on} = \frac{A^1{}_{on}}{A^1{}_{on} + A^2{}_{on} + A^2{}_{off}} \tag{7.10}$$

$$P^2{}_{on} = \frac{A^2{}_{on}}{A^1{}_{on} + A^2{}_{on} + A^2{}_{off}} \tag{7.11}$$

$$P^2{}_{off} = \frac{A^2{}_{off}}{A^1{}_{on} + A^2{}_{on} + A^2{}_{off}} \tag{7.12}$$

where $A^1{}_{on}$ is the average duration when only one channel is ON; $A^2{}_{on}$ is the average duration when both channels are ON; and $A^2{}_{off}$ is the average duration when both channels are OFF during a connection. let $\tau(n) = \frac{x(n)}{B}$ where $x(n)$ is the number of bits per frame as defined in (4.1), then $f_\tau(\tau) = B \cdot f_x(B\tau)$ and

$$A^1{}_{on} = \int_0^T \tau f_\tau(\tau)d\tau + \int_T^{2T} (2T-\tau)f_\tau(\tau)d\tau \quad (7.13)$$

$$A^2{}_{on} = \int_T^{2T} (\tau - T)f_\tau(\tau)d\tau \quad (7.14)$$

$$A^2{}_{off} = \int_0^T (T-\tau)f_\tau(\tau)d\tau . \quad (7.15)$$

T is the frame interval and is equal to $\frac{1}{F}$ of the video sequence, where F is the frame rate. Once $P^1{}_{on}$, $P^2{}_{on}$, and $P^2{}_{off}$ are computed as above, the next step is to quantify the aggregation of such multiple sources. We assume that the sources are homogeneous with similar traffic characteristics but are mutually independent with a phase uniformly distributed between $[0,T]$. Given that a total of j VBR video connections are active, the probability that c channels are concurrently ON can be determined as

$$P_{vd}(j,c) = \begin{cases} \sum_{i=0}^{\left\lfloor \frac{c}{2} \right\rfloor} \binom{j}{i} \binom{j-i}{c-2i} (P^2{}_{on})^i (P^1{}_{on})^{c-2i} (P^2{}_{off})^{j-i-(c-2i)} & \forall c \le 2j \\ 0 & \text{otherwise} \end{cases} \quad (7.16)$$

For example, $P_{vd}(j,3)$ = (the probability that 3 out of j connections have only 1 channel ON and remaining j-3 connections have both channels OFF) + (the probability that 1 out of j connections has 2 channels ON, 1 out of remaining j-1 connections have 1 channel ON, and the rest j-2 connections have both channels OFF). Also, the average number of channels that would be ON, given j active connections, is $2 \cdot j \cdot P^2_{on} + j \cdot P^1_{on}$.

The above analysis could be easily extended to video sources employing MPEG (Motion Pictures Experts Group) compression scheme as individual I, P and B frame series are considered to be AutoRegressive with Gaussian distribution [39, 13].

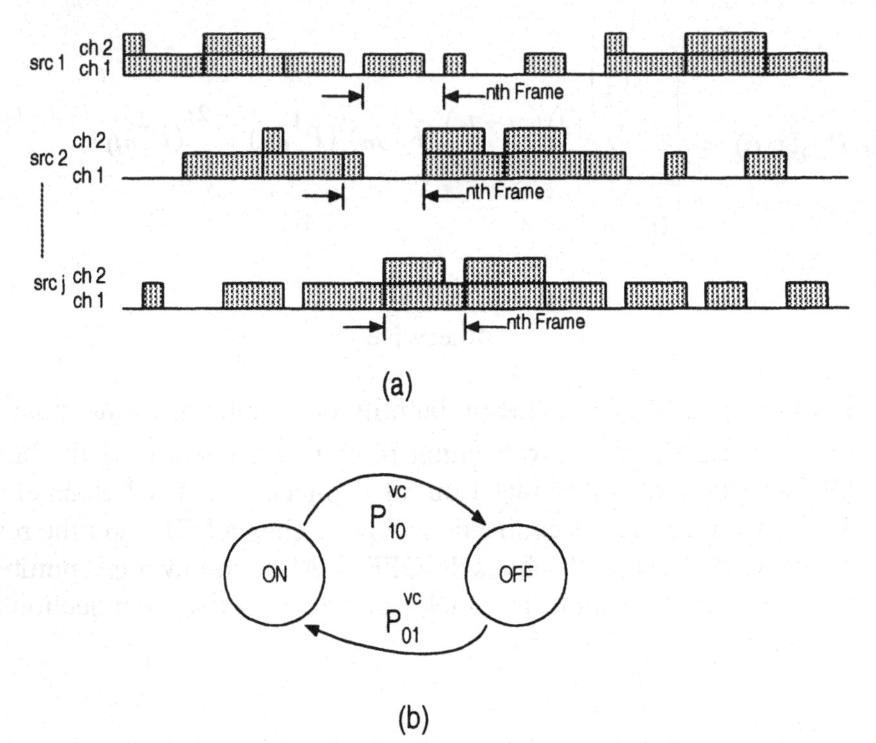

Figure. 7-1. Channel Activity Models (a) VBR Video Source (b) Voice Source.

b) Voice Sources

A voice source with silence detection is modeled as a conventional two state Markov process. Thus, given that j independent voice calls are active, the probability that c channels are simultaneously ON is given by

$$P_{vc}(j, c) = \begin{cases} \binom{j}{c}(p_{on})^c (p_{off})^{j-c} & \forall c \leq j \\ 0 & \text{otherwise} \end{cases} \tag{7.17}$$

where P_{on} and P_{off} are the probabilities that a voice channel is ON and OFF respectively. If P_{01} and P_{10} are the rate of transitions from ON to OFF and OFF to ON states respectively, as illustrated in Figure 7-1(b) then

$$p_{off} = \frac{p_{01}}{p_{01} + p_{10}} \text{ and } p_{on} = \frac{p_{10}}{p_{01} + p_{10}}. \quad (7.18)$$

c) CBR Sources

For CBR sources, given j active connections, the probability that c channels are simultaneously ON is given by

$$P_{cbr}(j, c) = \begin{cases} 1 & \text{if} \quad j = c \\ 0 & otherwise \end{cases}. \quad (7.19)$$

7.3.1.2 Statistical Multiplexing Gain

Consider a CDMA cell supporting traffic from I different services. A service s_i, where $i = 1..I$, may be a CBR service or one of the VBR services discussed above, with $P_i(j_i,c_i)$ being the probability that c_i channels are ON given j_i active connections of type s_i. We define $Pe(c)$ as the probability of error or bit error rate when any c channels are ON. Let Pe_i be the maximum tolerable bit error rate for s_i. Pe_i is a QoS parameter that is a direct measure of SINR requirements of service s_i, and is assumed to be known. Two services may have similar channel utilization characteristics but different maximum tolerable bit error rate, perhaps due to revenue considerations. Let C_i be the maximum number of channels, from any of the sources, that could be simultaneously ON while still assuring that the probability of error is less than or equal to Pe_i *i.e.*,

$$C_i = max\{c : Pe(c) \leq Pe_i\} \quad (7.20)$$

Then, assuming that each s_i call can utilize at most m_i channels, the maximum number of s_i only calls that could be admitted in the cell is simply $J_i = \left\lfloor \frac{C_i}{m_i} \right\rfloor$. However, the probability that all the active sources will be in the ON state at the same time could be negligible. Thus, taking the channel utilization models of the previous section into consideration, the maximum number of s_i only calls that could be admitted is now

$$J_i = max\left\{ j_i : \sum_{c > C_i} P_i(j_i, c_i) \leq \varepsilon_i \right\}. \tag{7.21}$$

$P_i(j_i, c_i)$ is the generalization of (7.16), (7.17) and (7.19) for any service s_i, which could be voice, video or a CBR service. In other words $P_i(j_i, c_i)$ is the probability that given a total of j_i connections of service s_i, only c_i channels are ON at the same time. The parameter ε_i is the allowed bit error rate tolerance *i.e.*, $Probability((Pe(c) \geq Pe_i) \leq \varepsilon_i)$. The effective capacity C_i for service s_i in the cell increases when ε_i>0, as more s_i only calls are allowed with some compromise with the bit error rate tolerance. Furthermore, let (j_1 , j_2 ,....., j_i,...., j_I) represent the number of connections from each service that could be active in the cell at a given time. A set of states that would assure the bit error rate requirements of service s_i, with tolerance ε_i, is given as:

$$A_i = \{(j_1, j_2, \ldots, j_i, \ldots, j_I) :$$

$$\left(\sum_{c_1 + c_2 + \ldots + c_i + \ldots + c_I > C_i} \prod_{i=1}^{I} P_i(j_i, c_i) \right) \leq \varepsilon_i \} \tag{7.22}$$

7.3.2 Cell Level QoS Evaluation

Given a particular channel access control policy, Pb_i is derived as follows. For simplicity, again, a two service case is assumed *i.e.*, $I = 2$. The analysis could be easily extended to a multi-service case.

7.3.2.1 Complete Sharing

The complete sharing implies uncontrolled access to the cell's capacity. The calls are served on a first-come-first-served basis. It is assumed that $Pe_1 \geq Pe_2$, $C_1 \geq C_2$ and $m_1 \leq m_2$. In the absence of any s_2 calls in the cell, higher number of s_1 calls could be accepted as a higher bit error rate is acceptable to s_1. If, on the other hand, an s_2 call is already in progress in the cell, a lower number of s_1 calls could be accepted to satisfy the bit error rate requirements of s_2. The set of possible acceptable states are illustrated in Figure 7-2, for both cases *i.e.*, with as well as without the bit error rate tolerance. A is the set of acceptable states and is defined as

$$A = (j_1, j_2) : \left(\sum_{c_1 + c_2 > C_2} \prod_{i=1}^{2} P_i(j_i, c_i) \right) \leq \varepsilon_2 \text{ or}$$

$$\left(j_2 = 0 \text{ and } \sum_{c_1 > C_1} P_1(j_1, c_1) \leq \varepsilon_1 \right) \tag{7.23}$$

If however ε_1=0 and ε_2=0, implying that no bit error rate tolerance is allowed then, A is as illustrated in Figure 7-3(a) and is given as

$$A = \{(j_1, j_2) : j_1 m_1 + j_2 m_2 \leq C_2 \text{ or } (j_2 = 0 \text{ and } j_1 m_1 \leq C_1)\} \tag{7.24}$$

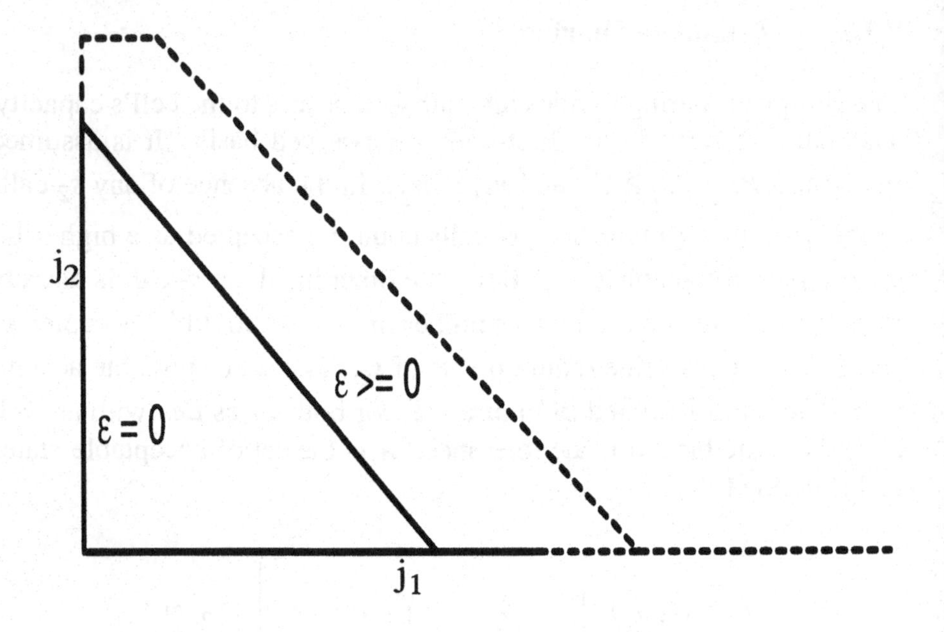

Figure. 7-2. Acceptable states for $\varepsilon = 0$ and $\varepsilon \geq 0$.

The acceptable region for both cases (with and without bit error rate tolerance) is coordinate convex [18]. Thus using the product-form solution of (5.5) and considering the case where ε_1=0 and ε_2=0 as in Figure 7-3(a), the blocking probabilities for the two services are given as

$$Pb_1 = P\left(\left\lfloor \frac{C_1}{m_1} \right\rfloor, 0\right) + \sum_{\left\{(j_1,j_2) \mid 1 \le j_2 \le \left\lfloor \frac{C_2}{m_2} \right\rfloor, j_1 = \left\lfloor \frac{C_2 - j_2 m_2}{m_1} \right\rfloor \right\}} P(j_1, j_2) \quad (7.25)$$

and

$$Pb_2 = \sum_{j_1 = \left\lfloor \frac{C_2}{m_1} \right\rfloor + 1}^{\left\lfloor \frac{C_1}{m_1} \right\rfloor} P(j_1, 0) + \sum_{\left\{(j_1,j_2) \mid 0 \le j_1 \le \left\lfloor \frac{C_2}{m_1} \right\rfloor, j_2 = \left\lfloor \frac{C_2 - j_1 m_1}{m_2} \right\rfloor \right\}} P(j_1, j_2). \quad (7.26)$$

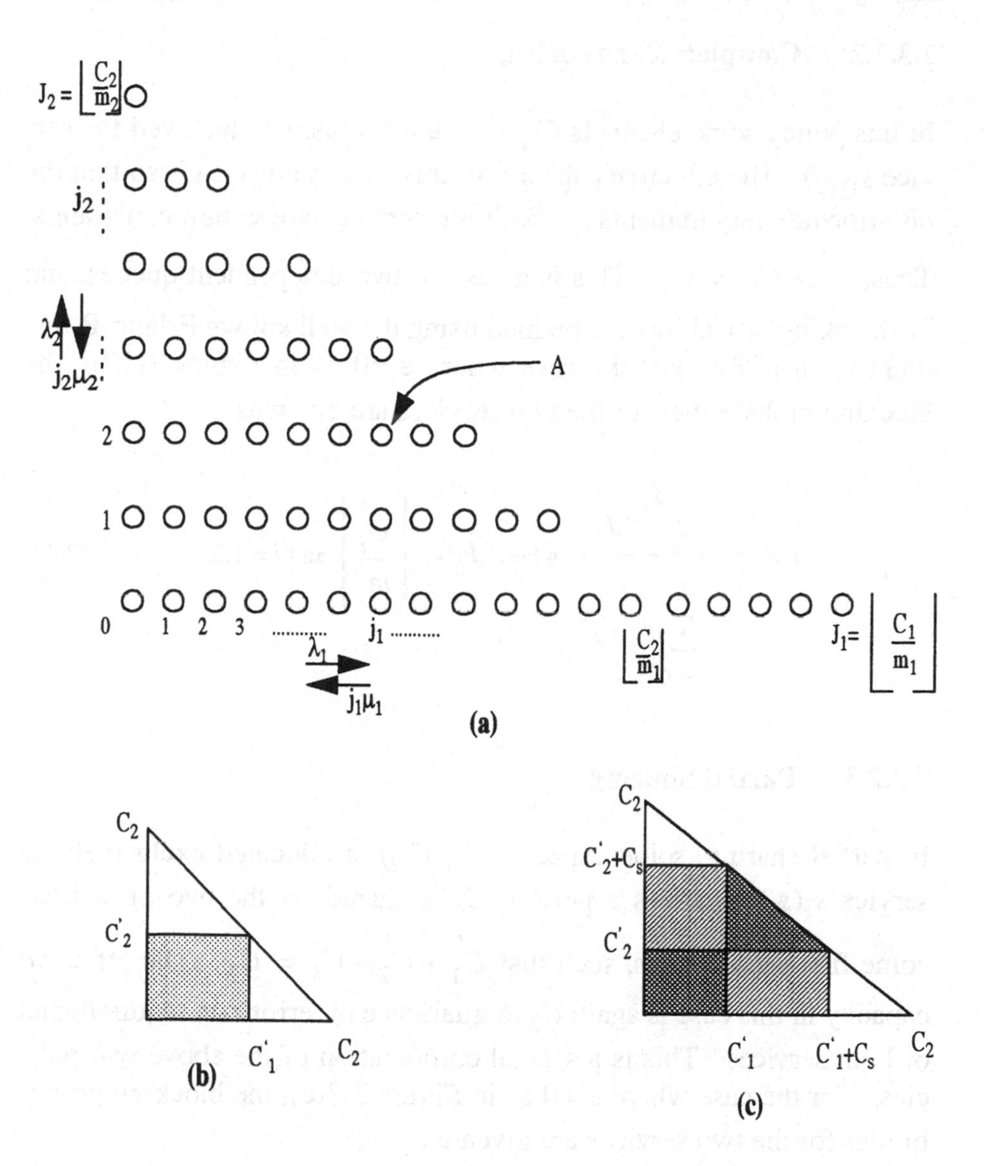

Figure. 7-3. Two Dimensional Markov chain when ε = 0 for (a) Complete Sharing (b) Complete Partitioning (c) Partial Sharing.

7.3.2.2 Complete Partitioning

In this policy some channels $C'_1(C'_2)$ are exclusively reserved for service $s_1(s_2)$. The effective capacity in this case is always C_2 so that the bit error rate requirements for both the services are within constraints. Thus, $C'_1 + C'_2 = C_2$. This is a case of two independent queues, and the blocking probability is obtained using the well known Erlang-B formula as in (5.8). For the case where ε_2=0 as in Figure 7-3(b), the blocking probabilities for the two services are given as

$$Pb_i = \frac{\rho_i^{J_i}/J_i}{\sum_{j=0}^{J_i} \rho_i^j/j!} \quad \text{where } J_i = \left\lfloor \frac{C'_i}{m_i} \right\rfloor \text{ and } i = 1,2. \tag{7.27}$$

7.3.2.3 Partial Sharing

In partial sharing, some capacity $C'_1(C'_2)$ is allocated exclusively to service $s_1(s_2)$, whereas a portion C_s is shared by the two on a first-come-first-served basis, such that $C'_1 + C'_2 + C_s = C_2$. The effective capacity in this case is again C_2 to guarantee bit error rate requirements of both services. This is a special combination of the above two policies. For the case where ε_2=0 as in Figure 7-3(c), the blocking probabilities for the two services are given as

$$Pb_1 = \sum_{\left\{(j_1, j_2) \mid 0 \le j_2 \le \left\lfloor \frac{C_s + C'_2}{m_2} \right\rfloor, j_1 = \min\left(\left\lfloor \frac{C_s + C'_1}{m_1} \right\rfloor, \left\lfloor \frac{C_2 - j_2 m_2}{m_1} \right\rfloor\right)\right\}} P(j_1, j_2) \tag{7.28}$$

and

$$Pb_2 = \sum_{\left\{(j_1,j_2) \mid 0 \le j_1 \le \left\lfloor \frac{C_s + C_1'}{m_1} \right\rfloor, j_2 = min\left(\left\lfloor \frac{C_s + C_2'}{m_2} \right\rfloor, \left\lfloor \frac{C_2 - j_1 m_1}{m_2} \right\rfloor\right)\right\}} P(j_1, j_2). \quad (7.29)$$

7.4 Results

In this section the applicability of the analysis conducted in the previous sections is demonstrated. Firstly, it is demonstrated that independent but identically distributed VBR video sources will have similar corresponding values for P^1_{on}, P^2_{on}, and P^2_{off}. This in turn implies that the statistical multiplexing gain, involving VBR video sources, can be estimated using the results of Section 7.3.1. Secondly, the performance of CS, CP and PS based channel access control policies is evaluated in terms of call level QoS parameters such as call blocking probability and call dropping probability. Also demonstrated is the conjecture that significant achievements in throughput could be realized by relaxing the bit error rate tolerance minimally and, thus, allowing statistical multiplexing.

A large ensemble of sample functions defined by (4.1) is generated. Each sample function represents a traffic trace from a compressed VBR video source, and has 5400 sample points, implying a 3 minute long video sequence (assuming F = 30 frames/sec). Each sample point in the sample function represents the size of the corresponding compressed frame in bits. Different values of the channel rate B are considered. For a frame size less than or equal to B/F bits, only single channel is assumed to be needed. On the other hand, for a frame size greater than B/F bits an additional channel is used in parallel to transmit the bits in excess of B/F. P^1_{on}, P^2_{on}, and P^2_{off} for each sample function are measured. The ensemble averages of the measured P^1_{on}, P^2_{on}, and P^2_{off} are listed in Table 7-1, for the specified B/F. The variance for each case is less than 0.001 which confirms the ergodicity of the sample functions as well as similarity in channel utilization models of independent but identically distributed VBR video sources. The proximity of P^1_{on}, P^2_{on}, and P^2_{off} computed numerically using (7.10),

(7.11), & (7.12), to the corresponding empirical values suggests that the parameters could be estimated using the analytical results of Section 7.3.1.

TABLE 7-1. Channel Utilization Model Parameters for VBR Video Sources.

B/F * (2.5 * 10^5)	**P^2on (empirical)**	**P^2on (numerical)**	**P^1on (empirical)**	**P^1on (numerical)**	**P^2off (empirical)**	**P^2off (numerical)**
0.55	0.147	0.144	0.669	0.664	0.182	0.191
0.60	0.099	0.097	0.682	0.675	0.217	0.226
0.65	0.067	0.064	0.681	0.673	0.250	0.262
0.70	0.044	0.041	0.671	0.661	0.284	0.297
0.75	0.027	0.026	0.649	0.642	0.322	0.331
0.80	0.017	0.016	0.630	0.619	0.352	0.364
0.85	0.010	0.009	0.604	0.594	0.384	0.396
0.90	0.005	0.005	0.579	0.568	0.415	0.426
0.95	0.003	0.003	0.552	0.542	0.443	0.454
1.00	0.002	0.002	0.529	0.517	0.469	0.480

A two dimensional CDMA cell grid with uniform traffic distribution across the area is considered to evaluate CS, CP and PS policies under the context of CDMA networks. Uniform traffic distribution is assumed for simplicity. For clarity, only a voice and a video service are considered. The call blocking probability as well as the call dropping probability of each service is computed using (7.8) and the formulations derived in Section 7.3.2 respectively. The impact of traffic intensity on these QoS parameters in the presence of the aforementioned channel access control policies is studied.

The effective capacities (C_1,C_2) for $\varepsilon_1=0$ & $\varepsilon_2=0$ are taken to be (48, 42) respectively. The other parameters such as (m_1,m_2), (λ_1,λ_2), and (μ_1,μ_2) are taken as (1,2), (120/121,1/121), and (1,0.05) respectively. The mean call holding time of a video call is, thus, assumed to be 20 times more than that of a voice call, whereas, the bandwidth requirement of a video call is twice that of a voice call. The call arrival rates (λ_1,λ_2) of both services are proportionally increased and the corresponding readings of the blocking probability are taken, that are presented in Figure 7-4. As expected, the CS policy exhibits unfairness. The blocking probability of wideband video service s_2, therefore, is substantially higher than that of narrowband voice service s_1. This difference, in fact, becomes more profound as the call arrival rates of both services increase. For CP policy (C'_1, C'_2) are taken to be (21,21) respectively. As shown in Figure 7-4, the blocking probability of s_2, in this case, improves significantly. The overall throughput, however, will deteriorate because of the bandwidth granularity. Figure 7-4 also depicts the PS case with (C'_1, C'_2) being (18,18) respectively, and C_s being 6 channels. Apparently, the PS policy can be used to optimize the call blocking probability and throughput characteristics, by fine tuning the shared and engineered bandwidth. Determination of accurate sizes of shared and engineered capacities is anything but trivial as the networks and services are always evolving. An adaptive update of channel allocations using the capacity optimization procedure outlined in Section 7.2 is, therefore, a potential solution. The system and traffic parameters used in the above mentioned results are chosen arbitrarily.

Figure 7-5 shows the dropping probabilities of the two services for the CS, CP and PS cases discussed above. As mentioned earlier, a uniform subscriber distribution and a random movement model is assumed. The values of (h_1, h_2) are assumed to be (0.09, 0.67) indicating that s_2 calls are more likely to go through handover because of higher average call holding time. The channel access control policies have the similar effect on the call dropping probability as on the call blocking probability, as expected from (7.8).

The enhancements in the effective capacity of a CDMA cell by relaxing the bit error rate tolerance are demonstrated through Figures 7-6 & 7-7. There, the parameters $(P^1_{on}, P^2_{on}, P^2_{off})$ for video sources are selected from Table 7-1 as (0.669,0.147,0.182), whereas, (P_{on}, P_{off}) for voice sources are chosen to be (0.4,0.6) being commonly used values for voice sources. By relaxing the bit error rate tolerance ε_1 as well as ε_2 to 10^{-5}, the effective capacities of each cell for the two services improve from (48,42) to (71,56). This results in a substantial drop in call blocking probability and call dropping probability for each of the two services, as is evident from Figures 7-6 & 7-7.

The advantage of using guard bands to handle handover calls, and thus reduce call dropping probability, is also demonstrated. A dropped call is less desirable than a blocked call. Some channels, therefore, may be explicitly reserved for handover calls to reduce the call dropping probability. A set of channels explicitly reserved in a cell to handle handover calls of a particular service is referred to here as a guard band. An optimum size of such a guard band could be determined by treating handover calls of a service as just another service. For clarity, consider for example that only one service is being offered; that λ_o is the rate at which the new calls originate in the cell; and that λ_h is the handover call arrival rate in that particular cell. Both are assumed to be Poisson, with channel service rate μ and capacity requirement of m channels per call. Consider that C_h channels are reserved for handover calls and the remaining cell capacity is shared by the handover calls as well as the calls originating in the cell. This situation can be treated as a special case of PS where $C^n{}_o + C^n{}_h + C^n{}_s \leq C^n$, with $C^n{}_o$ as zero. The blocking probability of handover arrivals is then directly obtained using the formulations derived in (7.28) & (7.29) for the PS case. The improvement in the call dropping probability with increase in the size of the guard band is demonstrated in Figure 7-8. A drop in the call dropping probability can be observed as the size of the guard band is increased. An optimum size of the guard band is determined using the capacity optimization procedure outlined in Section 7.2. Both, the handover arrival rate as well as the call origination rate of each service are specified as inputs to the procedure. The blocking probabilities of handover arrivals as well as originating calls are determined. Based on the call blocking probability and the call dropping probability requirements of each service, as well as, the overall revenue considerations, the optimum number of channels that are exclusively reserved for handover calls; the optimum number of channels that are shared by handover as well as originating calls of a service; and the optimum number of channels that are shared by different services, in a cell, are determined.

Finally, Figure 7-9 illustrates the change in network revenue as the reserved and shared capacities allocated to the two competing services *i.e.*, C'_1, C'_2 and C_s are varied. Again, a prominent global maxima exists, similar to the single link case of Chapter 6, which the capacity optimization procedure tries to achieve. In a multi-service network, however, the shape may not always be concave or convex; and there is always a possibility that the procedure may reach a local maxima and stop. In order to avoid this and to ensure that optimum global maxima is realized, the procedure could be run again, from a randomly selected initial state, as also suggested in Chapter 6.

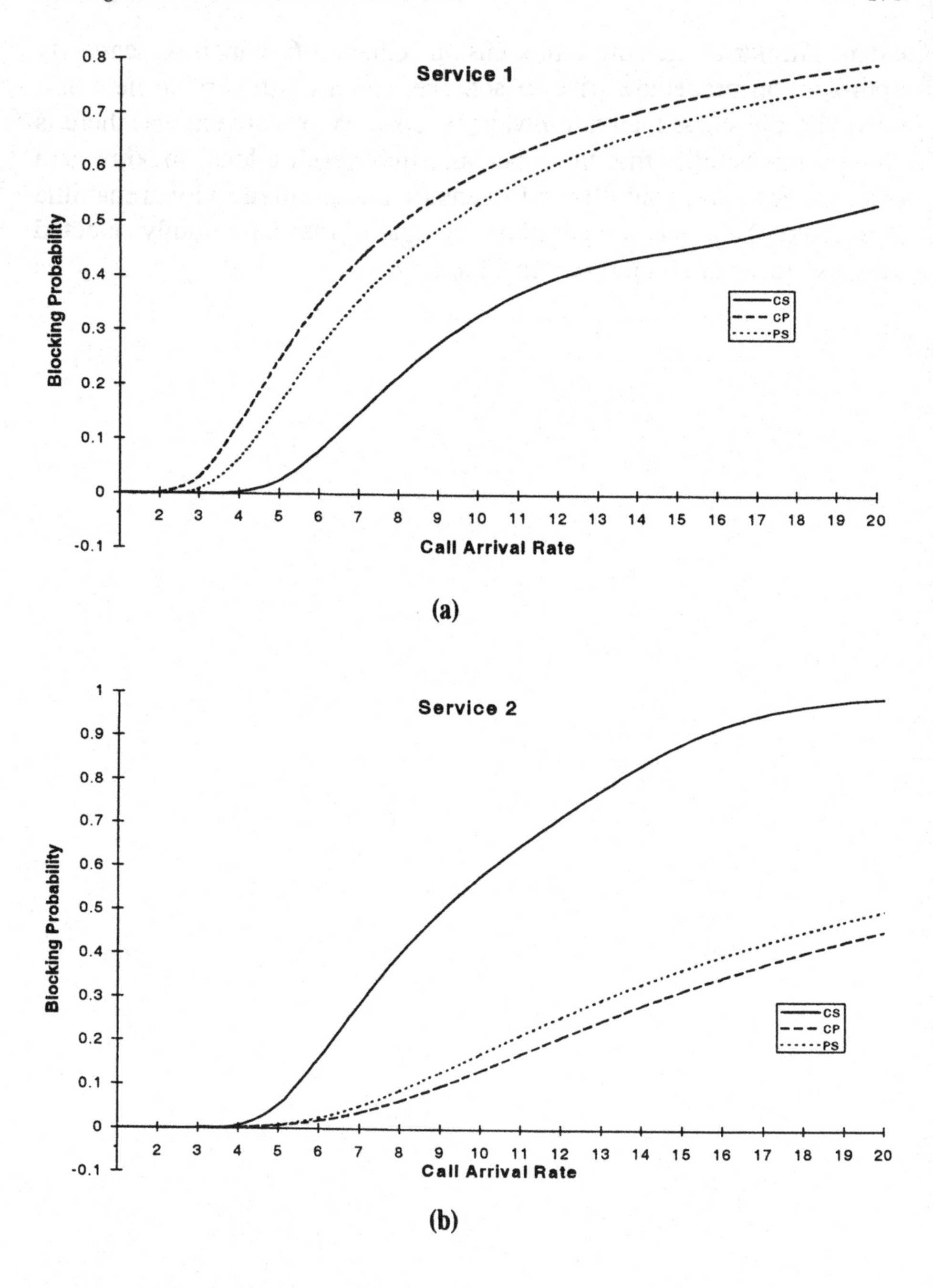

Figure. 7-4. Blocking Probability when $\varepsilon = 0$ for (a) Service 1 (b) Service 2.

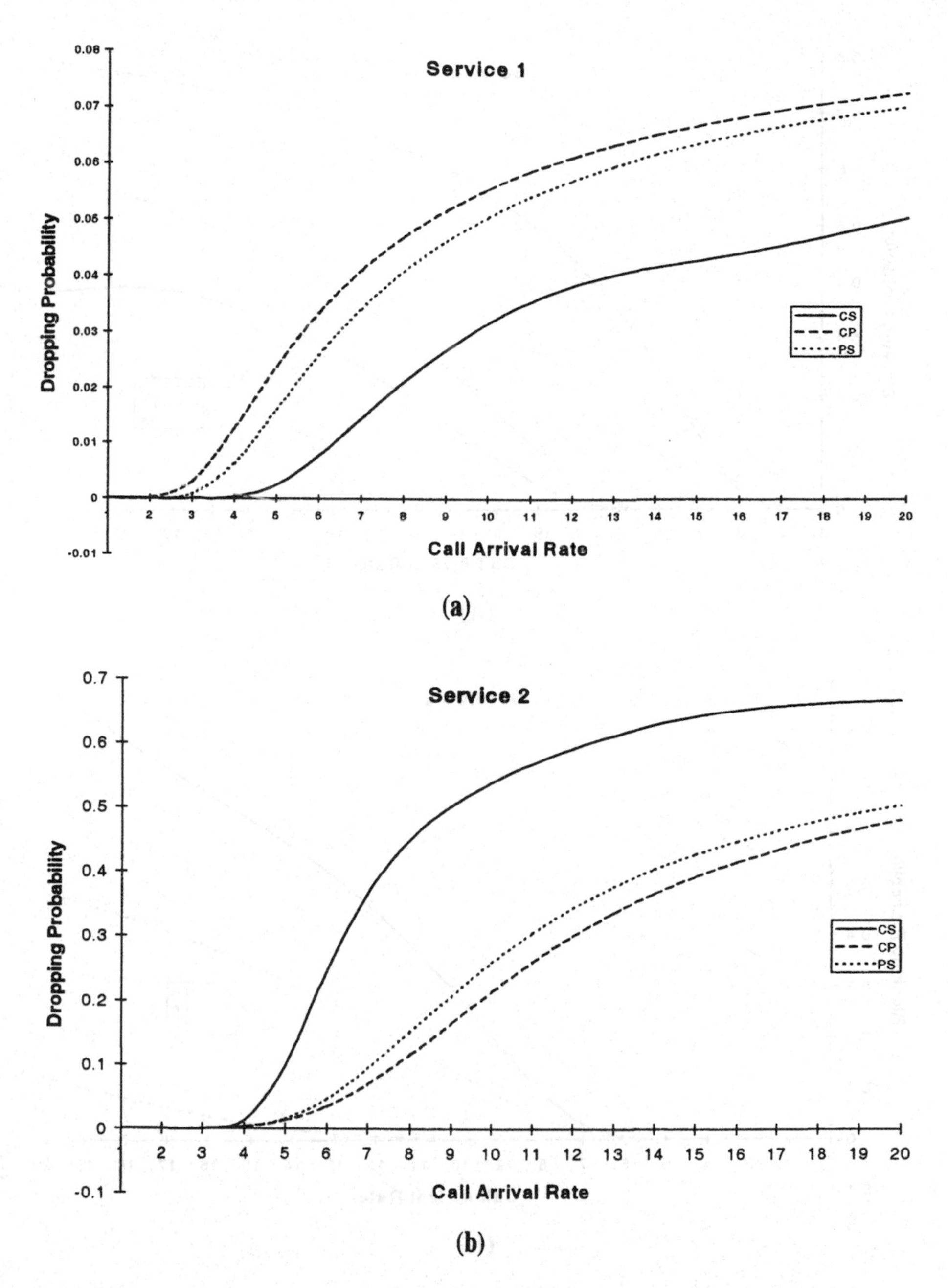

Figure. 7-5. Dropping Probability when ε = 0 for (a) Service 1 (b) Service 2.

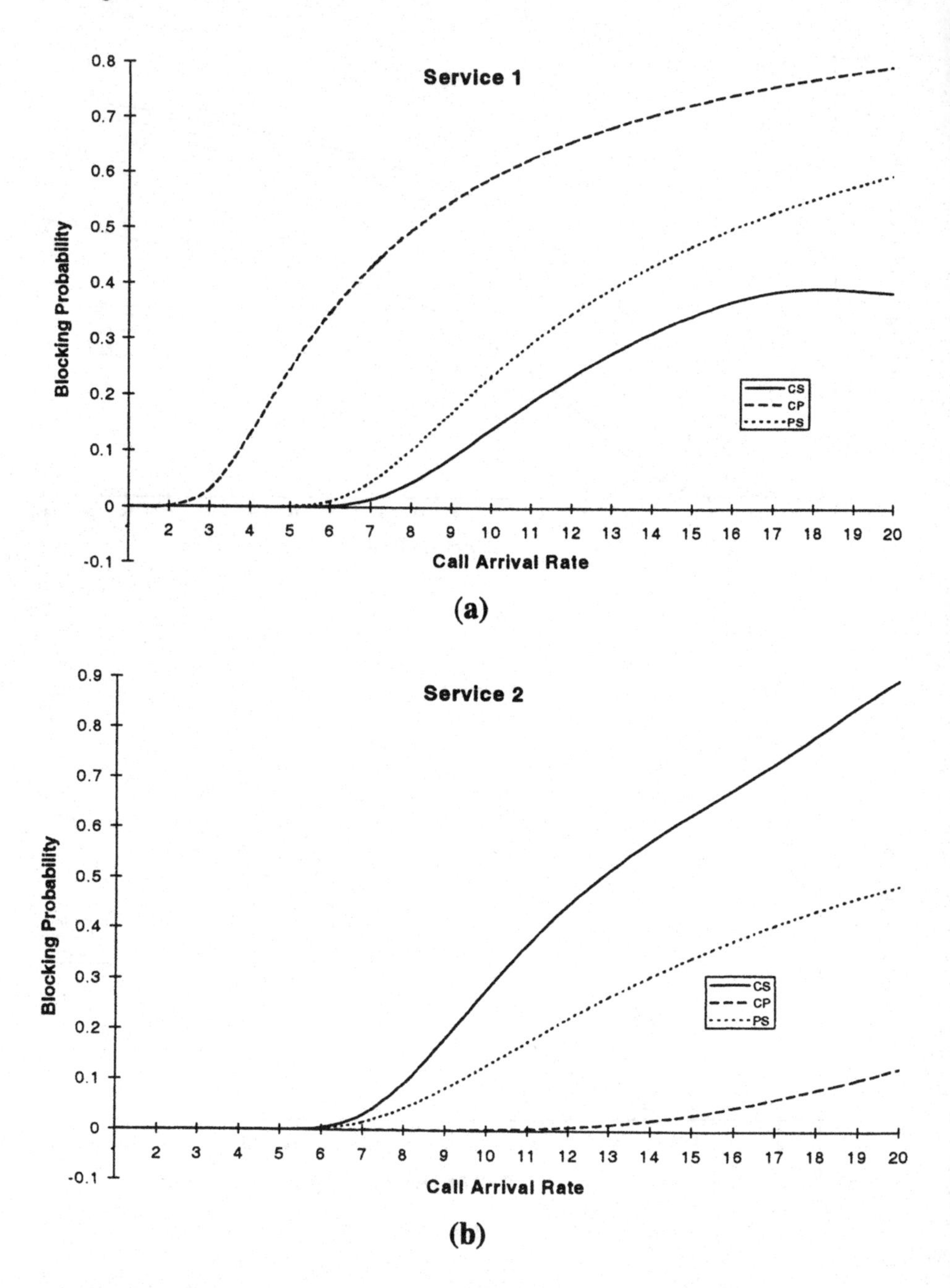

Figure. 7-6. Blocking Probability when $\varepsilon = 10^{-5}$ for (a) Service 1 (b) Service 2.

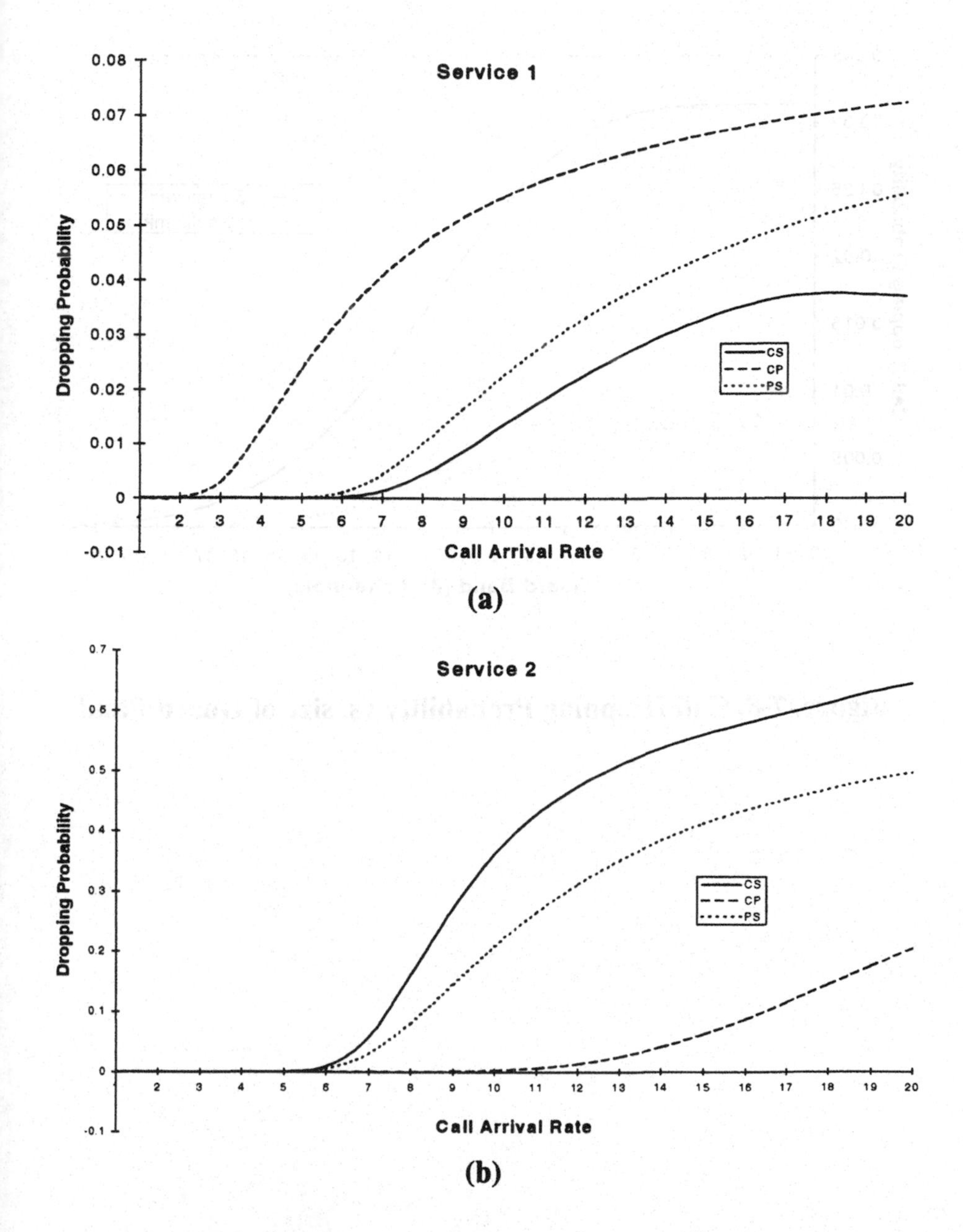

(a)

(b)

Figure. 7-7. Dropping Probability when $\varepsilon = 10^{-5}$ for (a) Service 1 (b) Service 2.

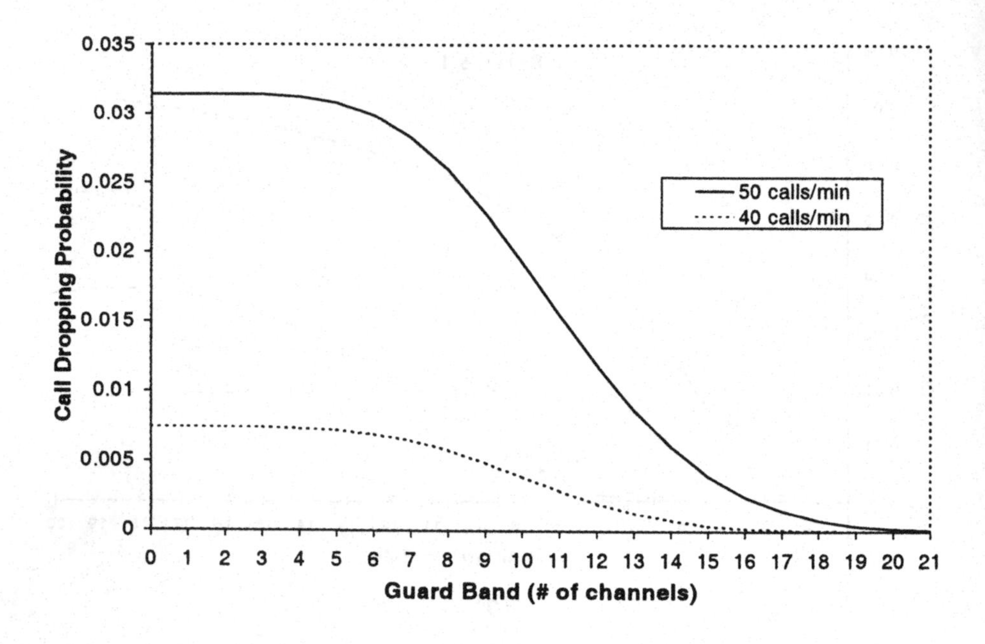

Figure. 7-8. Call Dropping Probability vs. size of Guard Band.

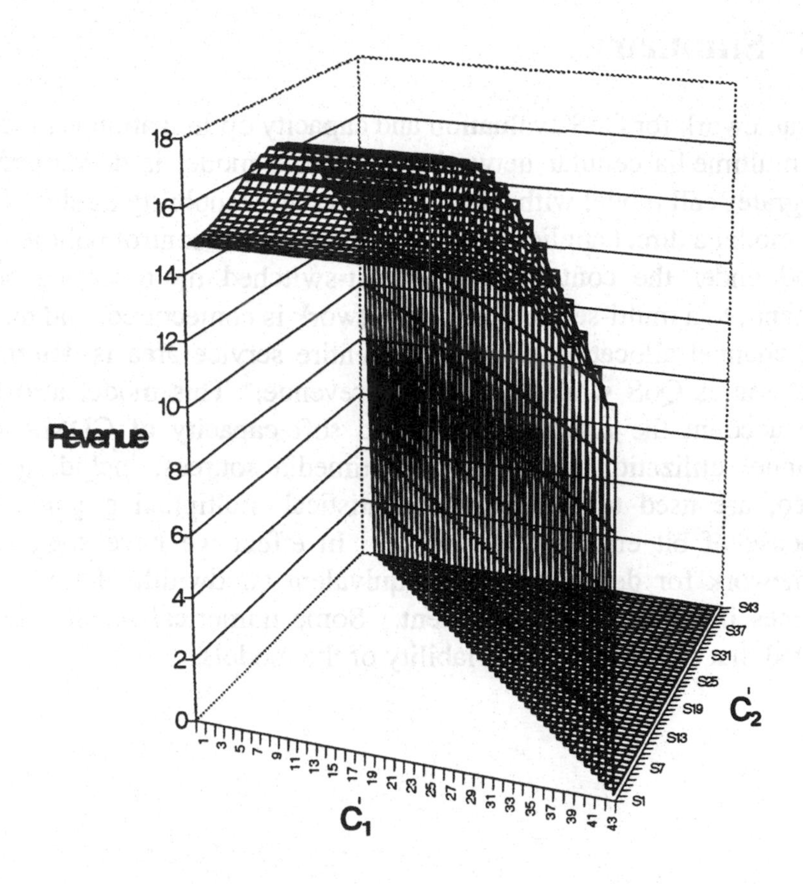

Figure. 7-9. Effect of reserving Channels for s_1 and s_2 on Network Revenue.

7.5 Summary

A framework for QoS evaluation and capacity optimization is proposed for multimedia cellular networks. A traffic model is developed that integrates call model with a Markov based user mobility model. Using this model a direct applicability of channel access control policies, analyzed under the context of a circuit-switched multi-service access demand, to a multi-service cellular network is conjectured; and an optimal channel allocation plan for the entire service area is determined that assures QoS while maximizing revenue. This model also takes into account the interference limited soft capacity of CDMA cells. Channel utilization models for multimedia sources, including VBR video, are used to estimate the statistical multiplexing gain at the expense of bit error rate tolerance. In effect we have suggested a framework for determining the equivalent bandwidth of multimedia sources in a CDMA environment. Some numerical results are presented that demonstrate the viability of the models.

Appendix A

Transient State Analysis of an (S+1) State Birth-Death Markov Process

This appendix derives (4.27) *i.e.*, the expression for the state transient probability distribution of an S+1 state birth-death Markov process. The differential-difference equation of (4.23) can be solved by making use of the probability generating function *P(z,t)* defined as

$$P(z, t) = \sum_{s=0}^{s=S} P_s(t) z^s \tag{A.1}$$

Multiplying both sides of (4.23) by z^s and summing over $s=0$ to $s=S$, substituting the initial conditions (4.24) and (4.25), and making use of the following properties of the probability generating functions *i.e.*,

$$\frac{d}{dt}\sum_{s=0}^{s=S} P_s(t) z^s = \frac{d}{dt}P(z, t)$$

and that $\frac{\partial}{\partial z}P(z, t)$ is equal to

$$\sum_{s=1}^{s=S} sP_s(t)z^{s-1} = \sum_{s=0}^{s=S-1} (s+1)P_{s+1}(t)z^s = \sum_{s=2}^{s=S+1} (s-1)P_{s-1}(t)z^{s-2}$$

produces

$$\frac{\partial}{\partial t}P(z,t) + [\lambda z^2 + (\mu - \lambda)z - \mu]\frac{\partial}{\partial z}P(z,t) = N\lambda(z-1)P(z,t) \quad (A.2)$$

(A.2) is of the form of Planar differential equations and can be solved by satisfying the following auxiliary system of differential equations.

$$\frac{dt}{1} = \frac{dz}{[\lambda z^2 + (\mu - \lambda)z - \mu]} = \frac{dP(z,t)}{N\lambda(z-1)P(z,t)} \quad (A.3)$$

By direct integration we get two solutions to (A.3)

$$\left(\frac{z-1}{\lambda z + \mu}\right)e^{-(\mu + \lambda)t} = C1 \quad (A.4)$$

$$\frac{P(z,t)}{(\lambda z + \mu)^S} = C2 \quad (A.5)$$

The constants *C1* and *C2* must be functionally related as C2 = g(C1) where g is an arbitrary function. Thus

$$P(z,t) = (\lambda z + \mu)^S g\left(\left(\frac{z-1}{\lambda z + \mu}\right)e^{-(\mu + \lambda)t}\right) \quad (A.6)$$

Consider an arbitrary initial condition that at time t=0 there were i sources active *i.e.*, $P_i(0) = 1$ where $0 \leq i \leq S$ and $P_s(0) = 0$ when $s \neq i$. In other words

$$P(z, 0) = \sum_{s=0}^{s=S} P_s(0) z^s = z^i \quad \text{(A.7)}$$

Substituting (A.7) in (A.6) gives

$$z^i = (\lambda z + \mu)^S g\left(\frac{z-1}{\lambda z + \mu}\right) \quad \text{(A.8)}$$

From (A.6) and (A.8) we get,

$$P(z, t) = \left(\frac{z + \frac{\mu}{\lambda}}{G + \frac{\mu}{\lambda}}\right)^S G^i \quad \text{(A.9)}$$

where

$$G = \frac{1 + \frac{\mu}{\lambda}\left(\frac{z-1}{z + \frac{\mu}{\lambda}}\right) e^{-(\mu+\lambda)t}}{1 - \left(\frac{z-1}{z + \frac{\mu}{\lambda}}\right) e^{-(\mu+\lambda)t}}$$

The state probabilities $P_s(t)$ can now be obtained by expanding the probability generating function in a power series, and since the probability generating function is being expanded, the coefficients of this power series then are the state probabilities $P_s(t)$. Considering the expression for $P(z,t)$ given by (A.9), the easiest way to expand $P(z,t)$ is to use Maclaurin's series, that is $P(z, t) = P(0, t) + zP'(0, t) + z^2P''(0, t) + z^3P'''(0, t) + \ldots$

Using the above expansion of $P(z,t)$, we obtain (2.27) *i.e.*, the expression for the state probabilities $P_s(t)$, which is

$$P_s(t) = \frac{1}{s!}\frac{\partial^s}{\partial z^s}P(z, t)\bigg|_{z=0} \tag{A.10}$$

Appendix B

Properties of Exponential Random Variables

In this appendix various properties of an exponential random variable are discussed and, thereafter, some of the results presented in the thesis are derived.

Let T_i, $1 \le i \le I$ be an exponential random variable with probability density function $f_{T_i}(t) = \mu_i e^{-\mu_i t}$ and, thus, cumulative density function $F_{T_i}(t) = 1 - e^{-\mu_i t}$. Assume that T1, T_2, ...T_I are independent. Let $T = \min\{T_1, T_2, ..., T_I\}$, then

$$P(T > t) = P(\min\{T_1, T_2, ..., T_I\} > t) =$$

$$P(T_1 > t, T_2 > t, ..., T_I > t) = P(T_1 > t) \cdot P(T_2 > t) \cdot ... \cdot P(T_I > t) =$$

$$(1 - F_{T_1}(t)) \cdot (1 - F_{T_2}(t)) \cdot ... \cdot (1 - F_{T_I}(t)) =$$

$$e^{-\mu_1 t} \cdot e^{-\mu_2 t} \cdot ... \cdot e^{-\mu_I t} = e^{-(\mu_1 + \mu_2 + ... + \mu_I)t} = e^{-\mu t}, \quad \text{(B.1)}$$

which implies that T is also an exponential random variable with rate $\mu = \mu_1 + \mu_2 + ... + \mu_I$.

Also,

$$P(T_2 > T_1) = \int_0^{\infty} (1 - F_{T_2}(t)) f_{T_1}(t) dt =$$

$$\int_0^{\infty} e^{-\mu_2 t} \cdot \mu_1 e^{-\mu_1 t} \cdot dt = \frac{\mu_1}{\mu_1 + \mu_2} \quad \text{(B.2)}$$

Similarly, for *I* such variables,

$$P(T_2, \ldots, T_I > T_1) = \frac{\mu_1}{\mu_1 + \mu_2 + \ldots + \mu_I}. \quad \text{(B.3)}$$

The above described properties are used in (5.11) and (7.4) as explained in the following sections.

B.1 Transition Rates in CD Policy

Consider a link where channel access control is based on a CD policy. Let the link be in a state (j_1, j_2) such that it jumps to $(j_1\text{-}1, j_2)$ if one of the j_1 calls of type s_1 completes; it jumps to $(j_1, j_2\text{-}1)$ if one of the j_2 calls of type s_2 completes; it jumps to $(j_1\text{+}1, j_2)$ if an s_1 call arrives; and it jumps to $(j_1\text{-}1, j_2\text{+}1)$ upon arrival of an s_2 call that preempts an active s_1 call. The link is modeled as a continuous-time multi-dimensional Markov process, therefore, the rate of transitions from (j_1, j_2) to $(j_1\text{-}1, j_2\text{+}1)$ is given by [5]

$$q_{(j_1, j_2), (j_1 - 1, j_2 + 1)} = \nu_{j_1, j_2} \cdot P_{(j_1, j_2), (j_1 - 1, j_2 + 1)}, \quad \text{(B.4)}$$

where $q_{(j_1, j_2), (j_1 - 1, j_2 + 1)}$ is the rate of transitions from (j_1, j_2) to $(j_1 - 1, j_2+1)$; ν_{j_1, j_2} is the rate at which the process makes transitions out of the state (j_1, j_2); and $P_{(j_1, j_2), (j_1 - 1, j_2 + 1)}$ is the probability that the process will jump to (j_1-1, j_2+1) when it moves out of the state (j_1, j_2). The process will move out of the state (j_1, j_2) if an active call completes or a new call arrives. As the call arrival processes of s_1 and s_2 are assumed to be Poisson (implying that the inter-arrival times are exponentially distributed) with parameters λ_1 and λ_2, respectively, from (B.1),

$$\nu_{j_1, j_2} = j_1\mu_1 + j_2\mu_2 + \lambda_1 + \lambda_2 . \tag{B.5}$$

Also, the process will jump to (j_1-1, j_2+1) only if an s_2 call arrives before the arrival of an s_1 call, and before one of the active calls completes. Thus, from (B.3),

$$P_{(j_1, j_2), (j_1 - 1, j_2 + 1)} = \frac{\lambda_2}{j_1\mu_1 + j_2\mu_2 + \lambda_1 + \lambda_2} . \tag{B.6}$$

From (B.4), (B.5) and (B.6)

$$q_{(j_1, j_2), (j_1 - 1, j_2 + 1)} = \lambda_2 , \tag{B.7}$$

as used in Section 5.5.

B.2 Handover Probability

Let T_h be the time spent by a mobile in a cell and T_d be the total call duration. Assuming that T_h and T_d are exponential random variables with parameters μ_h and μ_d, respectively, using (B.2), the probability that an active call goes through a handover before completion is given by

$$h = P(T_d > T_h) = \frac{\mu_h}{\mu_h + \mu_d}. \qquad \text{(B.8)}$$

B.3 Non-Exponential Call Durations

A system where some services may have non-exponentially distributed call durations can still be analyzed within the framework of a continuous-time Markov process using approximations. One such approximations based technique is the method of phases [57]. According to this method a non-exponentially distributed call duration is approximated as a sum, or a mixture, or a combined sum-mixture of independent, but not necessarily identical, exponentially distributed phases. The phases are not the actual phases during the call, but just the artifices introduced for the approximation purposes.

Consider a link with two competing services s_1 and s_2 with traffic parameters (λ_1, μ_1) and (λ_2, μ_2), respectively. Let the s_1 calls arrive according to a Poisson process with exponentially distributed call durations. Let s_2 calls arrive according to a Poisson process with arbitrary distributed call durations. Let T1 and T_2 be the random variables representing the call duration of an s_1 and s_2 call, respectively. Then, according to the method of phases

$$T_2 = T_{2_1} + T_{2_2} + \ldots + T_{2_n} \quad \text{(B.9)}$$

where $T_{2_1}, T_{2_2}, \ldots, T_{2_n}$ are independent, but not necessarily identical, exponential random variables with parameters $\mu_{2_1}, \mu_{2_2}, \ldots, \mu_{2_n}$, respectively. Then,

$$E\{T_2\} = \sum_{k=1}^{n} \frac{1}{\mu_{2_k}} \quad \text{(B.10)}$$

and

$$E\left\{(T_2)^2\right\} = \sum_{k=1}^{n} 2 \cdot \frac{1}{(\mu_{2_k})^2} \quad \text{(B.11)}$$

The other moments can also be fitted to better approximate the distribution of T_2 by judiciously choosing n and μ_{2_k}.

As an example, consider that the call duration of s_2 calls be approximated as a sum of two independent and exponentially distributed phases. This approximation implies that at any instant some of the s_2 calls could be in phase 1 while others in phase 2. Any accepted s_2 call first goes through phase 1 and thereafter transits to phase 2. The state of the link can now be represented as (j_1, j_{2_1}, j_{2_2}), where j_1 is the number of s_1 connections, and (j_{2_1}, j_{2_2}) are the number of s_2 connections in phase 1 and phase 2, respectively. The link now can be modeled as a multi-dimensional continuous-time Markov process. As an example, let the channel access control be based on the CD policy.

Consider that the link is in one of the droppable states as in Section B.1 *i.e.*, an arriving s_2 call will preempt an existing s_1 call. Based on the above approximation, the transition rates associated with this state are as follows:

$$(j_1, j_{2_1}, j_{2_2}) \rightarrow (j_1 + 1, j_{2_1}, j_{2_2}) \quad \text{at rate} \quad \lambda_1,$$

$$(j_1, j_{2_1}, j_{2_2}) \rightarrow (j_1 - 1, j_{2_1}, j_{2_2}) \quad \text{at rate} \quad j_1\mu_1,$$

$$(j_1, j_{2_1}, j_{2_2}) \rightarrow (j_1 - 1, j_{2_1} + 1, j_{2_2}) \quad \text{at rate} \quad \lambda_2,$$

$$(j_1, j_{2_1}, j_{2_2}) \rightarrow (j_1, j_{2_1} - 1, j_{2_2} + 1) \quad \text{at rate} \quad j_{2_1}\mu_{2_1}, \text{ and}$$

$$(j_1, j_{2_1}, j_{2_2}) \rightarrow (j_1, j_{2_1}, j_{2_2} - 1) \quad \text{at rate} \quad j_{2_2}\mu_{2_2},$$

provided $m_1 j_1 + m_2(j_{2_1} + j_{2_2}) \le C$, where C is the total number of channels available at the link; and m_1 and m_2 are the bandwidth requirements of s_1 and s_2 calls, respectively. Similarly, the transition rates of other valid states could be obtained, and, thereafter, the state probabilities could be determined by solving the global balance equations of the above process.

Appendix C

Capacity Optimization Procedure

In this appendix, the capacity optimization procedure, outlined in Chapter 7, is expressed using flow charts. The symbols used in the flow charts are as under:

- $i = 1..I$, is a service type s_i.
- $n = 1..N$ is a cell within the service area.
- $m = 1..M$, is a cell in a cluster (*i.e.*, a cell in the neighborhood of cell n).
- C_i^n is the number of channels reserved for a service s_i in a cell n.
- C_s^n is the number of channels shared by all the services in a cell n, on a first-come-first-served basis, or some other priority-based criteria such as CD or DA, as proposed in Chapter 5.
- $\left\{C_i^n, C_s^n\right\}$ represents the network-wide channel allocation plan *i.e.*, it is a set of C_i^n and C_s^n for all i and n.

- $\left\{ \mathrm{currC}_i^n, \mathrm{currC}_s^n \right\}$ represents a configuration obtained by perturbing the actual configuration by 1 BBU.

- $\left\{ \mathrm{saveC}_i^n, \mathrm{saveC}_s^n \right\}$ represents a potential configuration.

- *V*(.) is a measure of how far the network is from satisfying the call blocking probability and the call dropping probability constraints in the network. It is one of the objective functions that is minimized by the capacity optimization procedure. It is a function of the channel allocation plan and is determined as $V\left(\left\{ C_i^n, C_s^n \right\}\right) = \sum_i \max(0, Pb_i - MaxPb_i) + \sum_i \max(0, Pd_i - MaxPd_i)$ The parameters Pb_i and Pd_i are obtained using (7.1) and (7.6), respectively.

- *J* is the network revenue obtained using (7.3)

The overall functional blocks of the procedure are shown in Figure C-1. In Step 1, the objective is to minimize *V*(.) by perturbing the existing configuration by 1BBU and checking if it reduces *V*(.). The iterations of Step 1 are illustrated in Figure C-2. For clarity only one type of perturbations are illustrated in Figure C-2. An affected service may not only borrow channels from the shared pool as exclusive capacity, but also borrow channels as shared pool from another competing service, or borrow channels exclusively from another competing service. Only the configurations that improve *V*(.) are considered for implementation. In Step 2, the objective is to minimize *V*(.) by borrowing channels from neighboring cells either as shared or exclusive capacity, as illustrated in Figure C-3. Again, for clarity, only one type of perturbations are illustrated. If $V(.) = 0$, implying that the call blocking probabilities and the call dropping probabilities of all the offered services over the entire network are within the constraints then, in Step 3, attempts are made to maximize the revenue J, using similar perturbations as in Step 1 and Step 2. Only those configurations are considered that improve J while ensuring that $V(.)= 0$.

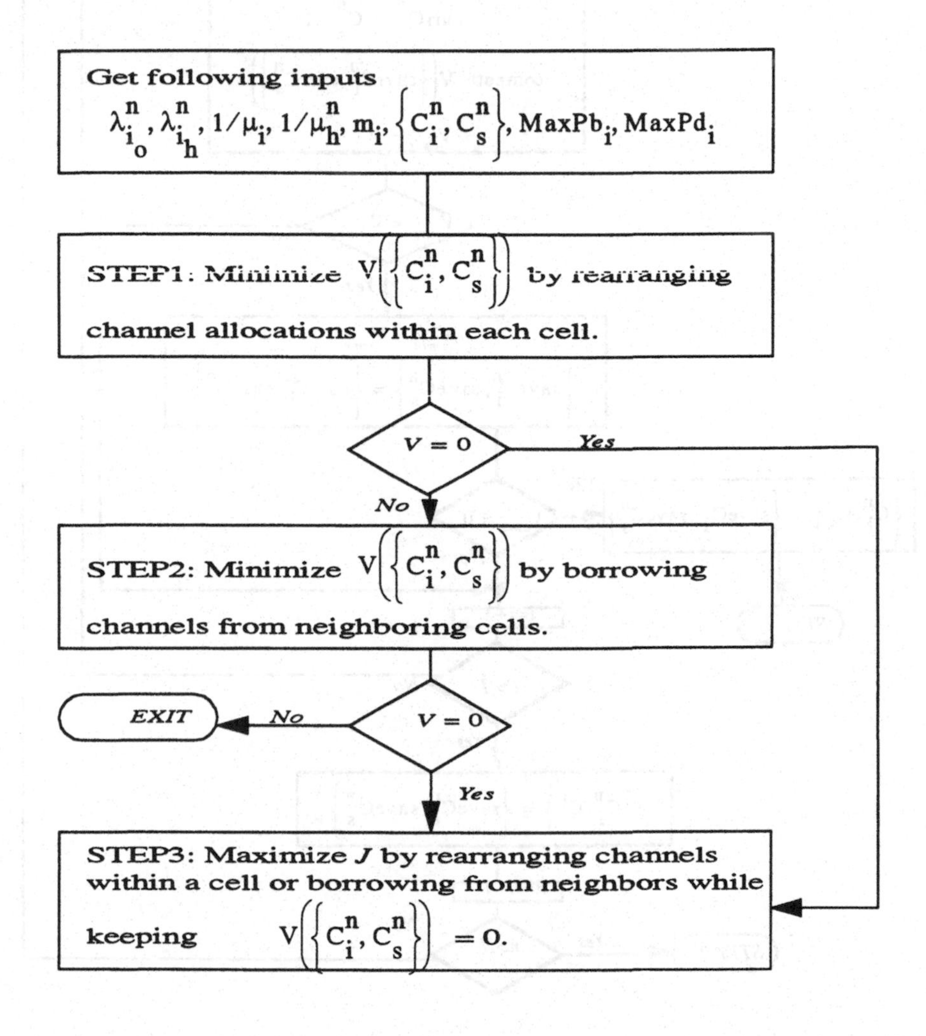

Figure. C-1. The Main Functional Blocks.

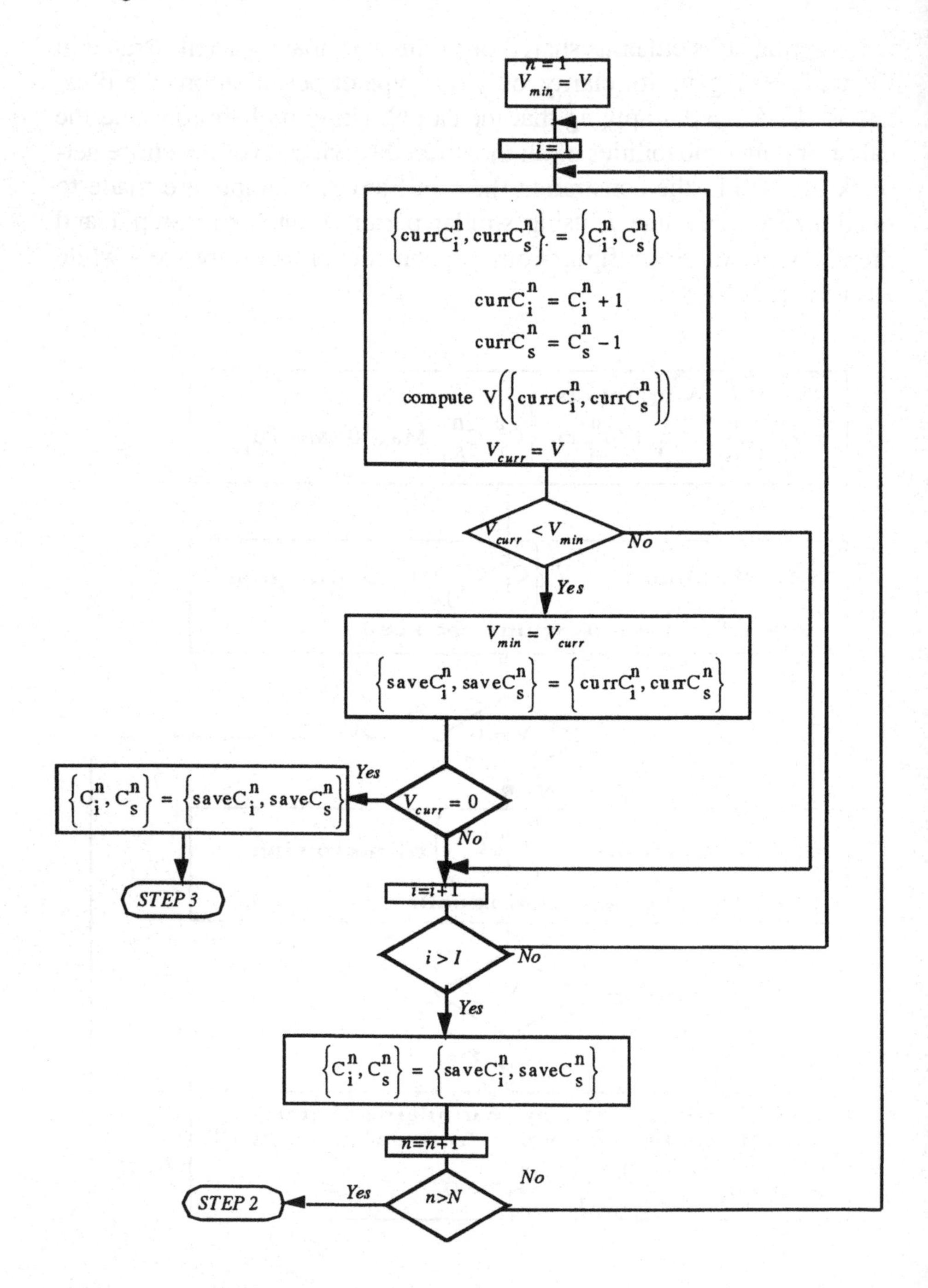

Figure. C-2. Iterations of Step 1.

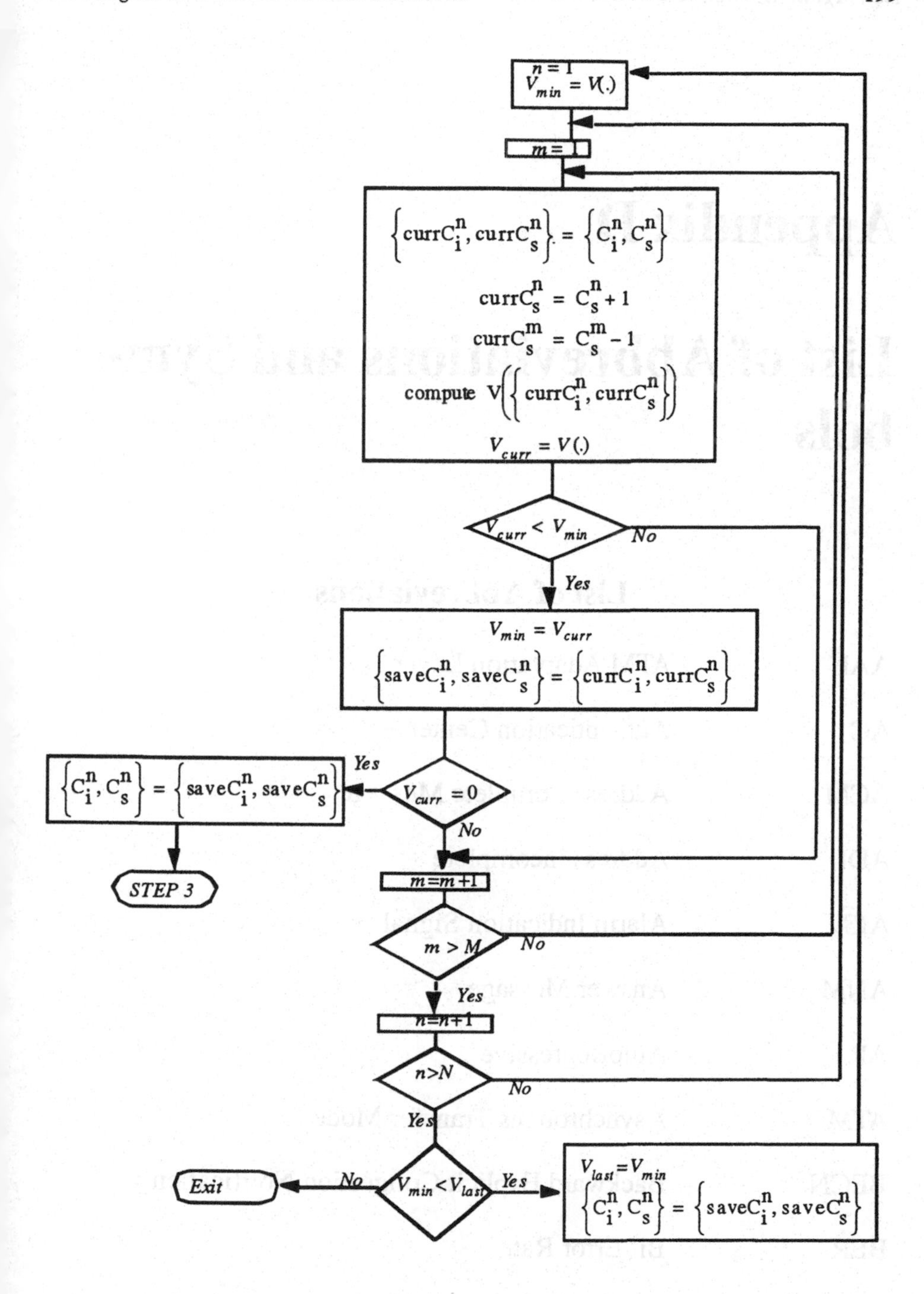

Figure. C-3. Iterations of Step 2.

Appendix D

List of Abbreviations and Symbols

List of Abbreviations

AAL	ATM Adaptation Layer
AC	Authentication Center
ACM	Address Complete Message
ADI	Address Incomplete
AIS	Alarm Indication Signal
ANM	Answer Message
AR	AutoRegressive
ATM	Asynchronous Transfer Mode
BECN	Backward Explicit Congestion Notification
BER	Bit Error Rate
BIP	Bit Interleaved Parity
B-ISDN	Broadband Integrated Services Digital Networks

BLA	Blocking Acknowledgment
BLO	Blocking Request
BSC	Base Station Controller
BSSAP	Base Station System Application Part
BSSMAP	Base Station System Managment Application Part
BTS	Base Trans-receive System
CAC	Call Admission Control
CBR	Constant Bit Rate
CDMA	Code Division Multiple Access
CDV	Cell Delay Variation
CFL	Call Failure
CLP	Cell Loss Priority
COO	Changeover Order message
CP	Complete Partitioning
CPE	Customer Premises Equipment
CRC	Cyclic Redundancy Check
CS	Convergence sublayer
CSU	Customer Service Unit
CTD	Cell Transfer Delay
CV	Coding Violations
DCN	Data Communication Network

DCT	Discrete Cosine Transform
DLCI	Data Link Connection Identifier
DPC	Destination Point Code
DPE	Data Processing Element
DQ	Dropped Calls Queued
DSE	Data Storage Element
DSS1	Digital Subscriber Signaling System No 1
DTAP	Direct Transfer Application Part
DUP	Data User Part
EB	Errored Block
ECO	Emergency Changeover Order message
EDC	Error Detection Code
EOC	Embedded Operations Channel
ES	Errored second
ESF	Extended Super Frame
ESR	Errored Second Ratio
FDMA	Frequency Division Multiple Access
FE	Framing Error
FEBE	Far End Block Error
FECN	Forward Explicit Congestion Notification
FEPR	Far-End Performance Report

FERF	Far-End Receive Failure
FISU	Fill In Signal Units (5 octets)
FSS	Failed Seconds or Failed Signal State
GRS	Group Reset Request
GS	Generic Services
GSM	Group Special Mobile or Global System for Mobile Communications
GTT	Global Title Translation
HBA	Hardware Failure Group Blocking Acknowledgement
HEC	Header Error Check
HGB	Hardware Failure Group Blocking Request
HLR	Home Location Registry
IAM	Initial Address Message
iid	Independent and Identically Distributed
INAP	Intelligent Network Application Protocol
JPEG	Joint Photographic Experts Group
LAN	Local Area Network
LAPF-Core	Link Access Procedure F-Core
LCD	Loss-Of-Cell Delineation
LIM	Link Interface Module
LMS	Least Mean Square

LOC	Loss of Continuity
LOF	Loss of Frame
LOP	Loss of Pointer
LOS	Loss Of Signal
LQ	Low Priority Calls Queued
LSSU	Link Status Signal Unit (6 or 7 octets)
LTE	Line Terminating Element
MAP	Mobile Application Part
MGB	Maintenance Group Blocking Request
MGU	Maintenance Group Unblocking Request
MOC	Mobile Originating Call
MPEG	Moving Picture Expert Group
MSC	Mobile Switching Center
MSU	Message Signal Unit (14-278 octets)
MTC	Mobile Terminating Call
MTP	Message Transfer Part
MUA	Maintenance Group Unblocking Acknowledgement
NE	Network Element
NMAP	Network Management Architecture Protocol
NNI	Network Node Interface
NRM	Network Resource Management

OAM&P	Operations, Administration, Maintenance, and Provisioning
OC	Optical Carrier
OCD	Out-of-Cell Delineation
OFS	Out-of-Frame Second
OOF	Out-of-Frame
OPC	Originating Point Code
OS	Operations System
OSI	Open Systems Interconnection
PCM	Plane Communication Manager
PDH	Plesiochronous Digital Hierarchy
pixel	Picture Element
PS	Partial Sharing
PSTN	Public-Switched Telephone Networks
PT	Payload-Type
PTE	Path Terminating Element
PVC	Permanent Virtual Circuit
QoS	Quality of Service
RAI	Remote Alarm Indication
REL	Release Message
RLC	Release Completed Message

RLS	Recursive Least Squares
RSC	Reset Circuit
S	Switching
SAR	Segmentation and Reassembly Sublayer
SCCP	Signaling Connection Control Part
SCP	Service Control Point
SDDCH	Standalone Dedicated Control Channel
SDH	Synchronous Digital Hierarchy
SEF	Severely Errored Frame
SES	Severely errored seconds
SESR	Severely Errored Second Ratio
SIF	Signaling Information Field
SINR	Signal to Interference and Noise Ratio
SIO	Service Indicator Octet
SLS	Signaling Link Selection
SMDS	Switched Multimegabit Data Service
SMS	Short Message Service
SNR	Signal to Noise Ratio
SONET	Synchronous Optical Network
SS7	Signaling System 7
SSP	Service Switching Point

STE	Section Terminating Element
STP	Signalling Transfer Point
STS	Synchronous Transport Signal
SU	Signaling Unit
SVC	Switched Virtual Circuits
T	Transmission
TC	Transmission Convergence
TCAP	Transaction Capabilities Application Part
TCH	Traffic Channel
TDMA	Time Division Multiple Access
TMN	Telecommunications Management Network
TUP	Telephone User Part
UAS	Unavailable Seconds
UBA	Unblocking Acknowledgement
UBL	Unblocking Request
UNI	User-Network Interface
UPC/NPC	Usage/Network Parameter Control
US	User Services
VBR	Variable Bit Rate
VC	Virtual Connection
VCI	Virtual Connection Id

VLR	Visitor Location Registry
VP	Virtual Path
VPI	Virtual Path Id
VT	Virtual Tributary

List of Symbols

$\boldsymbol{a}$	A bold lowercase italic implies a vector.
$\boldsymbol{A}$	A bold uppercase italic implies a Matrix.
f_a	The probability density function of a random variable a.
F_a	The cumulative density function of a random variable a.
$E\{a\}$	The expectation of a random process a.
$r_{aa}(t)$	The autocorrelation function of a random process a.
$\boldsymbol{r}_{aa}$	The autocorrelation vector of a random process a.
$\boldsymbol{R}_{aa}$	The autocorrelation matrix of a random process a.
$s(t)$	The number of sources in an active state at time t.
F	The frame rate of a video sequence.
T	The frame length or inter-frame interval.
w	An observation interval.
W	The number of observation windows during a frame length.
$c(j)$	The cell or packet counts in the jth observation window.
$x(j)$	Transmission rate at the multiplexor output during j^{th} observation interval.
$\hat{x}(j)$	An estimate of $x(j)$.
$x_i(n)$	The discrete-time inter-frame bit-rate process of ith source

$\tilde{x}_i(t)$ The continuous-time inter-frame bit-rate process of ith source.

$x_i(t)$ The continuous-time inter-frame bit-rate process of ith source assuming a random phase.

f The vector of least mean-square filter coefficients.

λ The average call arrival rate.

μ The average call duration or call holding time.

$P_s(t)$ The probability that s sources are active at time t.

s_i The ith service type.

m_i The number of channels occupied by a call from i^{th} service type.

C Maximum number of channels in a wired link or a wireless link (i.e. cell).

C_i The number of channels allocated to the ith service type in a link.

C_i The number of channels shared by all competing services in a link.

j_i The number of connections from the ith service type in a link.

$\boldsymbol{j}$ The link state i.e. number of connections from each service type.

A A set of acceptable link states $\boldsymbol{j}$.

$P(\boldsymbol{j})$ The probability that the link is in state $\boldsymbol{j}$.

λ_i The average call arrival rate to a link of ith service type.

μ_i The average channel occupancy time of a call from i^{th} service type in a link.

Pb_i The network-wide call blocking probability of ith service type.

Pd_i The network-wide call dropping probability of ith service type.

γ_i The network-wide throughput of ith service type in calls/sec.

ρ_i The traffic intensity of ith service type in a link.

R_i The average revenue generated by the ith service type.

h_i^n The handover probability of the ith service type from cell n.

$P_i^{n_1 n_2}$ The probability that a call of the ith service type is handed off to the neighboring cell n_2, from cell n_1

$P_i(j_i, c_i)$ The probability that given a total of j_i connections of service s_i, only c_i channels are ON at the same time

ε_i The bit error rate tolerance.

Pe_i The maximum tolerable bit error rate for the ith service type calls in a CDMA cell.

$Pe(c)$ The bit error rate if c channels are simultaneously ON in a CDMA cell.

$P_{on(off)}$ The probability that a voice channel is ON(OFF).

P^1_{on} The probability that only one channel of a VBR source is ON.

$P^2_{on(off)}$ The probability that both channels of a VBR source are ON(OFF).

A^1_{on} The average fraction of time when only one channel of a VBR source is ON.

$A^2_{on(off)}$ The average fraction of time when both channels of a VBR source are ON(OFF).

References

[1] A. Adas, "Supporting Real Time VBR Video Using Dynamic Reservation Based on Linear Prediction," in *Proc. IEEE INFOCOM'96*, San Francisco, California, 1996, pp. 1476-1483.

[2] J. Aein, "A Multi-User-Class, Blocked-Calls-Cleared, Demand Access Model," *IEEE Trans. Communication,* vol. COM-26, no. 3, pp. 378-385, March 1978.

[3] N. Anerousis and A. Lazar, "Virtual Path Control for ATM Networks with Call Level Quality of Service Guarantees," *IEEE Trans. Networking,* vol. 6, no. 2, pp. 222-236, April 1998.

[4] Bellcore, "Synchronous Optical Network (SONET) Transport Systems: Common Generic Criteria," TA-TSY-000253.

[5] D. Bertsekas and R. Gallager. *Data Networks*. 2nd Edition, Prentice Hall, 1992.

[6] R. Bouchard and C. Pyers, "Use of Gravity Model for Describing Urban Travel," *Hwy. Res. Rec.*, vol. 88, pp. 1-43, 1965.

[7] P. Clarkson. *Optimal and Adaptive Signal Processing*. CRC Press, 1993.

[8] S. Chong, S. Li and J. Ghosh, "Predictive Dynamic Bandwidth Allocation for Efficient Transport of Real-Time VBR Video over ATM," *IEEE J. Select. Areas Commun.*, vol. 13, no. 1, pp. 12-23, January1995

[9] A. Viterbi and A. Viterby, "Erlang Capacity of a Power Controlled CDMA System," *IEEE J. Select. Areas Commun.*, vol.11, no. 6, pp. 892-900, Aug. 1993.

[10] C. Comaniciu and N. Mandayam, "Delta Modulation Based Prediction for Access Control in Integrated Voice/Data CDMA Systems," *IEEE J. Select. Areas Commun.*, vol. 18, no. 1, pp. 112-122, Jan. 2000.

[11] C. Chan, "Performance Analysis of a Multi-Class CDMA Network," in *Proc. IEEE Veh. Technol. Conf.'99*, Amsterdam, the Netherlands, 1999, pp. 411-415.

[12] S. Dixit and P. Skelly, "Video Traffic Smoothing and ATM Multiplexer Performance," in *Proc. IEEE GLOBECOM'91*, Phoenix, Arizona, 1991, pp. 239-243.

[13] N. Doulamis, A. Doulamis, G. Konstantoulakis and G. Stassinopoulos, "Performance Models for Multiplexed VBR MPEG Video Sources" in *Proc. IEEE ICC'97,* Montreal, Canada, 1997, pp. 856-861.

[14] B. Epstein and M. Schwartz, "Reservation Strategies for Multi-Media Traffic in a Wireless Environment," *IEEE Veh. Technol. Conf'95*, Chicago, Illinois, 1995, pp. 165-169.

[15] J. Evans and D. Everitt, "Effective Bandwidth-Based Admission Control for Multiservice CDMA Cellular Networks," *IEEE Trans. Veh. Technol.*, vol. 48, no. 1, pp. 36-46, Jan. 1999.

[16] E. Ferrara and B. Widrow, "The Time-Sequenced Adaptive Filter," *IEEE Trans. ASSP,* vol. ASSP-29, no. 3, pp. 679-683, June 1981.

[17] W. Gardner (Editor). *Cyclostationarity in Communications and Signal Processing*. IEEE Press, 1994.

[18] J. Kaufman, "Blocking in a Shared Resource Environment," *IEEE Trans. Communication*, vol. COM-29, no. 10, pp. 1474-1481, Oct. 1981.

[19] D. Hong and S. Rappaport, "Traffic Model and Performance Analysis for Cellular Mobile Radio Telephone Systems with Prioritized and Nonprioritized Handoff Procedures," *IEEE Trans. Veh. Technol.,* vol. VT-35, no. 3, pp. 77-92, Aug. 1986.

[20] G. Meempat and M. Sundareshan, "Optimal Channel Allocation Policies for Access Control of Circuit-Switched Traffic in ISDN Environments," *IEEE Trans. Communication,* vol. 41, no. 2, pp. 338-350, Feb. 1993.

[21] S. Jung and J. Meditch, "Adaptive Prediction and Smoothing of MPEG Video in ATM Networks," in *Proc. IEEE ICC'95*, Seattle, Washington, 1995, pp. 832-836.

[22] T. Kabasawa, T. Watanabe and M. Sengoku, "Transient Characteristics of Mobile Communication Traffic in a Band Shaped Service Area," *ICICE Trans. Fundamentals,* vol. E76-A, no. 6., June 1993.

[23] S. Kheradpir, A. Gersht and W. Stinson "Performance Management in SONET-Based Multi-Service Networks," *IEEE GLOBECOM'91*, Phoenix, Arizona, 1991, pp. 1406-1411.

[24] D. Knisely, S. Kumar, S. Laha and S. Nanda, "Evolution of Wireless Data services: IS-95 to cdma2000," *IEEE Comm. Mag.*, vol. 36, no. 10, pp. 140-149, Oct. 1998.

[25] A. Kolarov, A. Atai, and J. Hui, "Application of Kalman Filter in High-Speed Networks," in *Proc. IEEE GLOBECOM'94,* San Francisco, California, 1994, pp. 624-628.

[26] D. Lam, D. Cox and J. Widom, "Teletraffic Modeling for Personal Communication services," *IEEE Comm. Mag.,* vol. 35, no. 2, pp. 79-87, Feb. 1997.

[27] A. Lambardo, G. Morabito, S. Palazzo and G. Schembra, "Intra-GOP Modeling of MPEG Video Traffic," in *Proc. IEEE ICC'98,* Atlanta, Georgia, 1998, pp. 563-567.

[28] P. Larijani, R. Hafez and I. Lambadaris, "Two Level Access Control Strategy for Multimedia CDMA,"in Proc. *IEEE ICC'98,* Atlanta, Georgia, 1998, pp. 487-492.

[29] K. Leung, W. Massey and W. Whitt, "Traffic Models for Wireless Communication Networks," *IEEE J. Select. Areas Commun.,* vol. 12, no. 8, pp. 1353-1364, Oct. 1994.

[30] H. Lev-Ari, "Adaptive RLS Filtering Under the Cyclo-Stationary Regime," in *Proc. IEEE ICASSP'98*, pp. 2185-2188, 1998.

[31] B. Maglaris, D. Anastassiou, P. Sen, G. Karlsson, and J. Robbins, "Performance Models of Statistical Multiplexing in Packet Video Communications," *IEEE Trans. Communication*, vol. 36, no. 7, pp. 834-844, July 1988.

[32] J. Markoulidakis, G. Lyberopoulos and M. Anagnostou, "Traffic model for Third Generation Cellular Mobile Telecommunication Systems," *Wireless Networks,* pp. 389-400, vol. 4, 1998.

[33] J. Chandramohan, "An Analytic Multiservice Performance Model for a Digital Link with a Wide Class of Bandwidth Reservation Strategies," *IEEE J. Select. Areas Commun.*, vol. 9, no. 2, pp. 220-225, Feb. 1991.

[34] S. Nanda, K. Balachandran and S. Kumar, "Adaptation Techniques in Wireless Packet Data Service," *IEEE Comm. Mag.*, vol. 38, no. 1, pp. 54-64, Jan. 2000.

[35] A. Papoulis. *Probability, Random Variables, and Stochastic Processes.* McGraw-Hill, 1984.

[36] Y. Qian, D. Tipper and D. Medhi, "A Nonstationary Analysis of Bandwidth Access Control Schemes for Heterogeneous Traffic in B-ISDN," *IEEE INFOCOM'96*, San Francisco, California, 1996, pp. 730-737.

[37] T. Randhawa and R.. Hardy, "Application of AR Based Model in Proactive Management of VBR Traffic," in *Proc. IEEE ICATM'98*, Colmar, France, 1998, pp. 234-241.

[38] T. Randhawa and R. Hardy, "Estimation and Prediction of VBR Traffic in High-Speed Networks using LMS Filters," in *Proc. IEEE ICC'98*, Atlanta, Georgia, 1998, pp. 253-258.

[39] T. Randhawa and R. Hardy, "ProActive Management of MPEG Traffic in High-Speed Networks using Time-Sequenced RLS Filters," *IEEE ICATM'99*, Colmar, France, 1999, pp. 253-258.

[40] T. Randhawa, K. Moir and R. Hardy, "Traffic Measurement Based Bandwidth Dimensioning of Broadband Networks," *IEEE/IFIP NOMS'2000*, Honolulu, Hawaii, 2000, pp. 307-320.

[41] T. Randhawa and R. Hardy, "Performance Analysis of Bandwidth Partitioning in Multi-Service Broadband Networks," *IEEE* ATM*'2000*, Heidelberg, Germany, 2000, session D15.2.

[42] T. Randhawa and R. Hardy, "Performance Analysis of Multi-Service Wireless Networks with Mobile Users," *IEEE WCNC'2000*, Chicago, Illinois, session S18.5.

[43] R. Rodrigues, G. Mateus, and A. Loureiro, "On the Design and Capacity Planning of a Wireless Local Area Network," *IEEE/IFIP NOMS'2000*, Honolulu, Hawaii, 2000, pp. 335-347.

[44] O. Rose, "Statistical Properties of MPEG Video Traffic and their Impact on Traffic Modeling in ATM Systems," University of Wurzburg, Institute of Computer Sciences, Report No. 101, February 1995.

[45] M. Schwartz. *Broadband Integrated Networks*. Prentice Hall, 1996.

[46] D. Shen and C. Ji, "Admission Control of Multimedia Traffic for Third Generation CDMA Network," *IEEE INFOCOM'2000*, Tel Aviv, Israel, 2000, pp. 1077-1086, Jan. 2000.

[47] P. Skelly, M. Schwartz, and S. Dixit, "A Histogram-Based Model for Video Traffic Behavior in an ATM Multiplexer," *IEEE/ACM Trans. Networking*, vol. 1, no. 4, pp. 446-458, August 1993.

[48] P. Skelly, G. Pacifici, and M. Schwartz, "A Cell and Burst Level Control Framework for Integrated Video and Image Traffic," in *Proc. IEEE GLOBECOM'93*, pp. 334-341, 1993.

[49] R. Thomas, H. Gilbert and G. Mazziotto, "Influence of the Movement of the Mobile Station on the Performance of a Radio Cellular Network," *3rd Nordic Seminar*, paper 9.4, Sept. 1988.

[50] P. Tsingotjidis, J. Hayes, and H. Kim, "Estimation and Prediction Approach to Congestion Control in ATM Networks," in *Proc. IEEE GLOBECOM'94*, San Francisco, California, 1994, pp. 1785-1789.

[51] W. Verbiest, L. Pinnoo, and B. Voeten, "The Impact of the ATM Concept on Video Coding," *IEEE J. Select. Areas Commun.*, vol. 6, no. 9, pp. 1623-1632, December 1988.

[52] X. Wang, S. Jung and J. Meditch, "Dynamic Bandwidth Allocation for VBR Video Traffic Using Adaptive Wavelet Prediction," in *Proc. IEEE ICC'98*, Atlanta, Georgia, 1998, pp. 549-553.

[53] H. Xie and S. Kuek, "Priority Handoff Analysis," *IEEE Veh. Technol. Conf'93*, Secaucus, New Jersey, 1993, pp. 855-858.

[54] S. Yang. *CDMA RF System Engineering*. Artech House Publishers, 1998.

[55] W. Yang and E. Geraniotis, "Admission Policies for Integrated Voice and Data Traffic in CDMA Packet Radio Networks," *IEEE J. Select. Areas Commun.*, vol. 12, no. 4, pp. 654-664, May 1994.

[56] E. Yu and C. Chen, "Traffic Prediction using Neural Networks," in *Proc. IEEE GLOBECOM'93*, Houston, Texas, 1993, pp. 991-995.

[57] M. Neuts. *Matrix Geometric Solutions in Stochastic Models*. The John Hopkins University Press, 1981.

[58] ITU-T Recommendation M.3400, "TMN Management Functions".

[59] ITU-T Recommendation G.784, "Synchronous Digital Hierarchy (SDH) Management".

[60] ITU-T Recommendation M.3200, "TMN Management Services: Overview".

[61] ITU-T Recommendation G.774, "Synchronous Digital Hierarchy (SDH) Management Information Model for the Network Element view"

[62] ITU-T Recommendation G.826, "Error Performance Parameters and Objectives for International, Constant Bit Rate Digital Paths at or above the Primary Rate"

[63] ITU-T Recommendation G.821, "Error Performance of an International Digital Connection Forming Part of an ISDN".

[64] ITU-T Recommendation G.712, "Transmission Performance Characteristics of Pulse Code Modulation".

[65] ITU-T Recommendation I.321, "B-ISDN Protocol Reference Model and its Application".

[66] ITU-T Recommendation I.361, "B-ISDN ATM Layer Specification".

[67] ITU-T Recommendation I.311, "B-ISDN General Network Aspects".

[68] ITU-T Recommendation I.356, "B-ISDN ATM Layer Cell Transfer Performance".

[69] ITU-T Recommendation I.371, "Traffic Control and Congestion Control in B-ISDN".

[70] ITU-T Recommendation I.122, "Framework for Providing Additional Packet Mode Bearer Services".

[71] ITU-T Recommendation I.233, "Frame Mode Bearer Services - General Structure and Service Capabilities".

[72] ITU-T Recommendation I.370, "Frame Mode Bearer Services - Congestion Management".

[73] ITU-T Recommendation Q.922, "ISDN Data Link Layer Specification for Frame Mode Bearer Services".

[74] ITU-T Recommendation M.20, "Maintenance Philosophy for Telecommunication Networks".

[75] ITU-T Recommendation M.2100, "Performance Limits for bringing-into-service and Maintenance of International PDH Paths, Sections and Transmission Systems".

[76] ITU-T Recommendation M.2120, "Digital Path, Section and Transmission System Fault Detection and Localization Procedures".

[77] ITU-T Specification Series Q.700-Q.709, "Message Transfer Part".

[78] GSM Specification 04.08, "Mobile Radio Interface Layer 3 Specification".

[79] GSM Specification 03.02, "Network Architecture".

[80] GSM Specification 08.08, "Mobile-services Switching Centre - Base Station System (MSC - BSS) interface; Layer 3 specification".

[81] GSM Specification 03.60, "General Packet Radio Service (GPRS); Service description; Stage 2".

[82] GSM Specification 04.64, "General Packet Radio Service (GPRS); Mobile Station - Serving GPRS Support Node (MS-SGSN) Logical Link Control (LLC) layer specification".

[83] GSM Specification 08.18, " General Packet Radio Service (GPRS); Base Station System (BSS) - Serving GPRS Support Node (SGSN); BSS GPRS Protocol (BSSGP)".

[84] ANSI T1.105-1988, "Digital Hierarchy-Optical Rates and Formats Specifications".

[85] S. Akers, S. Donn, S. Gregg and S. Ravikumar, "Uses of In-Service Monitoring Information to Estimate Customer Service Quality", *IEEE GLOBECOMM* 1989, pp. 1856-1859.

[86] B. Bates. Introduction to T1/T3. Artech House, 1992.

[87] L. Blazek, "Performance Monitoring of Fiber Optics Systems," *IEEE GLOBECOMM* 1985, pp. 1367-1369.

[88] S. Farkouh, "Managing ATM-based Broadband Networks," *IEEE Communications Magazine*, May 1993, pp. 82-86.

[89] D. Gaiti and G. Pujolle, "Performance Management Issues in ATM Networks: Traffic and Congestion Control," *IEEE/ACM Transactions on Networking*, April 1996, pp249-257.

[90] J. Gruber, "Performance and Fault Management Functions for the Maintenance of SONET/SDH and ATM Transport Networks," *IEEE GLOBECOMM* 1993, pp. 1308-1314.

[91] Rony Holter, "SONET: A Network Management Viewpoint", *IEEE LCS*, Nov. 1990, pp. 4-13.

[92] J. Jakubson., "Managing SONET Networks," *IEEE LTS*, Nov. 1991, pp. 5-13

[93] R. Mallon and S. Ravikumar, "Detection of Bursty Error Conditions through Analysis of Performance Information," *IEEE GLOBECOMM* 1989, pp. 2020-2024.

[94] T.S. Randhawa, S. Hawthorne, K. Moir and R.H.S. Hardy, "An Integrated Capacity Manager for Multi-Media Cellular Networks," *IEEE IM* 2001, pp. 257-270.

[95] R. Sasisekharan, Y-K Hsu and D. Simon, "SCOUT: An Approach to Automating Diagnoses of Faults in Large Scale Networks", *IEEE GLOBECOMM* 1993, pp. 212-216.

[96] R. Sasisekharan, V. Seshadri, and S. M. Weiss. Proactive Network Maintenance Using Machine Learning.

[97] P. Smith and M. Shafi, "The impact of G.826 on the performance of Transport Systems," *IEEE/ACM Transactions on Networking*, pp. 604-614, Aug. 1996.

[98] S. Weiss and C. Kulikowski. Computer Systems That Learn: Classification and Prediction Methods from Statistics, Neural Nets, Machine Learning, and Expert Systems. Morgan Kaufmann, 1991.

Subject Index